LOW SALINITY AND ENGINEERED WATER INJECTION FOR SANDSTONE AND CARBONATE RESERVOIRS

LOW SALINITY AND ENGINEERED WATER INJECTION FOR SANDSTONE AND CARBONATE RESERVOIRS

EMAD WALID AL SHALABI
Khalifa University of Science and Technology, The Petroleum Institute, Abu Dhabi, UAE

KAMY SEPEHRNOORI
The University of Texas at Austin, Texas, USA

ELSEVIER

Gulf Professional Publishing
An imprint of Elsevier

Gulf Professional Publishing is an imprint of Elsevier
50 Hampshire Street, 5th Floor, Cambridge, MA 02139, United States
The Boulevard, Langford Lane, Kidlington, Oxford, OX5 1GB, United Kingdom

Library of Congress Cataloging-in-Publication Data
A catalog record for this book is available from the Library of Congress

British Library Cataloguing-in-Publication Data
A catalogue record for this book is available from the British Library

ISBN: 978-0-12-813604-1

For Information on all Gulf Professional Publishing publications
visit our website at https://www.elsevier.com/books-and-journals

Publishing Director: Joe Hayton
Senior Acquisition Editor: Katie Hammon
Editorial Project Manager: Katie Chan
Production Project Manager: Anusha Sambamoorthy
Cover Designer: Mark Rogers

Typeset by MPS Limited, Chennai, India

DEDICATION

To my beloved parents Walid and Fatin, dear brother Eyad, and lovely sisters Nancy, Nadia, and Noor, for their endless love, patience, inspiration, support, and encouragements.

Emad Walid Al Shalabi

To my brother Dr. Darius S. Noori.

Kamy Sepehrnoori

CONTENTS

LIST OF FIGURES

LIST OF TABLES

CHAPTER ONE

Introduction to Enhanced Oil Recovery Processes

Content

Different recovery mechanisms are involved during the life of an oil reservoir, including primary, secondary, and tertiary mechanisms. Primary recovery includes oil recovery by natural drive mechanism, including solution gas, water influx, gas cap drives, and gravity drainage. The conventional primary recovery ranges from 3% original oil in place (OOIP) with the help of the expansion of undersaturated oil to 15% OOIP with solution gas drive. In the secondary recovery phase, different process are used to raise or maintain reservoir pressure, such as gas or water injection. The presence of an active water drive or a gas cap drive boosts the recovery significantly to about 50% or more of the OOIP by maintaining the reservoir pressure via gas or water injection (Guerithault and Economides, 2001; Lake, 1989).

Most of the average pressure of oil reservoirs is depleted during the primary and secondary recovery phases. As a result of this pressure depletion, a large fraction of the OOIP is left behind in the reservoir. Different methods are suggested to recover the remaining oil in economic and environmental approaches (Dandona and Morse, 1972). For the tertiary recovery phase, enhanced oil recovery (EOR) methods are used to improve the oil recovery beyond primary and secondary recoveries in an economic way under certain market and technology conditions. The EOR is the oil recovery by injection of fluids that normally do not exist in the reservoir, which excludes pressure maintenance or waterflooding. This definition does not restrict the application of EOR to a particular phase (primary, secondary, or tertiary) (Lake, 1989). Different EOR techniques are being used, including solvents such as miscible and immiscible gas flooding (hydrocarbon, carbon dioxide, or nitrogen), chemical flooding (surfactant, polymer, or alkaline), thermal methods (steam flooding, cyclic steam

Low Salinity and Engineered Water Injection for Sandstone and Carbonate Reservoirs.
DOI: http://dx.doi.org/10.1016/B978-0-12-813604-1.00001-8

flooding, or in-situ combustion), and others [microbial, low salinity/engineered water injection (LSWI/EWI), or acoustic].

Improved oil recovery (IOR) is another term that has been used interchangeably or even replaced the EOR term. IOR refers to any process that improves oil recovery; hence, this definition includes EOR processes as well as other practices such as waterflooding, pressure maintenance, infill drilling, and multilateral wells. Fig. 1.1 shows the different recovery mechanisms during the life of an oil reservoir.

Waterflooding has been considered as the most commonly used secondary oil recovery technique since 1865. Recently, a tertiary effect of this technique has been observed depending on the composition and salinity of the injected water. LSWI/EWI is one of the emerging IOR techniques for wettability alteration in both sandstone and carbonate reservoirs. The popularity of this technique is due to its high efficiency in displacing light-to-medium gravity crude oils, ease of injection into oil-bearing formations, availability and affordability of water, and lower capital and operating costs involved. The latter advantages lead to favorable economics compared to other IOR/EOR methods.

The LSWI/EWI IOR technique is also known in the literature as LoSal by BP, Smart WaterFlood by Saudi Aramco, Designer Waterflood by

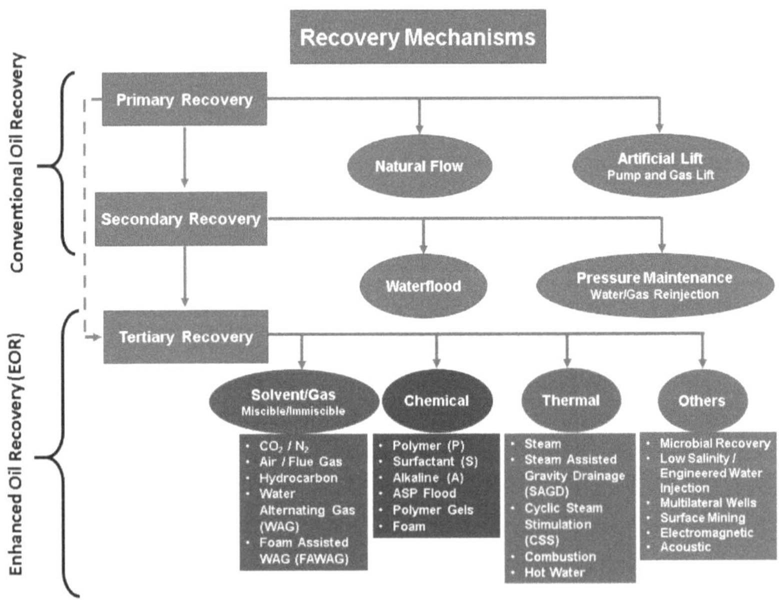

Figure 1.1 Oil recovery mechanisms during the life of an oil reservoir.

Shell, and Advanced Ion Management (AIM^{SM}) by ExxonMobil. Several studies have been conducted on LSWI at laboratory scale and to a limited extent at field scale. Most studies have confirmed a positive response to low salinity injection, which is translated into additional oil recovery in both secondary and tertiary injection modes. Wettability alteration is believed to be the main reason behind incremental oil recovery due to LSWI; however, some other mechanisms are suggested as well, such as dissolution and fine migration processes. Nevertheless, work is progressing on understanding chemical interactions among crude oil/brine/rock (COBR) in the system.

Few LSWI modeling studies have been performed so far especially on carbonate rocks compared to sandstone rocks. One of the reasons for this reluctance to investigate the effect of LSWI on carbonate rocks is the extensive research done on sandstone rocks, which concluded that the presence of clay is the main reason for wettability alteration. Moreover, the complex chemical interactions between COBR and the heterogeneity of carbonate rocks make it difficult to predict the extent of additional oil recovery as a result of LSWI. Other reasons involve the mystery of the chemical mechanism behind the oil increase with low salinity injection and the discrepancy in some of the published results regarding the effect of low salinity compared to the seawater injection effect.

The only concern with the low salinity waterflooding technique is water sourcing and water disposal. There are two main methods for seawater desalination: thermal-based and membrane-based. The thermal-based methods include multistage flash distillation, multi-effect distillation, thermo-compression distillation, and mechanical vapor compression. The thermal-based method is based on heating the saline feed water and collecting the condensed vapor from the distillation process. On the other hand, the membrane-based method includes reverse osmosis (RO) and nanofiltration (NF). The latter methods are pressure driven upon which the pressure is applied to force the saline feed water through a membrane and hence separation happens while the selective salts remain. Usually, membrane-based methods are favored on thermal-based methods, especially offshore due to space limitation and energy/steam requirements. The RO method is a tight membrane with pore sizes less than or equal to 0.0005 μm upon which the produced water is just fresh water with no monovalent or divalent ions. NF membranes are looser compared to RO with pore sizes ranging from 0.05 to 0.005 μm and the product water is rich in monovalent ions (Yousef and Ayirala, 2014).

RO desalination is the most common membrane-based method for water desalination; however, clay swelling and reservoir souring are two main related problems. A novel method suggested by Ayirala et al. (2010) overcomes the previously mentioned problems by using NF step and RO; it is known as designer water desaturation process. Several patents have been proposed for membrane-based desalination method in which RO alone or a series configuration of NF/RO is used. Nevertheless, Yousef and Ayirala (2014) highlighted that the proposed configuration just works for low salinity waterflooding in Sandstone. Moreover, they proposed a novel water ionic composition optimization technique upon which NF and RO membrane processes are used in parallel configuration. The latter configuration generates multiple product water streams to cover the entire range of ionic content and variation, which are suitable for both sandstone and carbonate rocks. Yousef and Ayirala stressed on the importance of the rejected NF and rejected RO streams, which are important in case of LSWI/EWIs in carbonate rocks.

Dang et al. (2013) presented a concise review of the current understanding of LSWI mechanisms, modeling and numerical simulation, LSWI pilot tests, and hybrid LSWI projects with focus on sandstone rocks. Moreover, Sheng (2014) provided a review on LSWI in sandstones including history of LSWI, laboratory and field observations, mechanisms, and simulation work. Also, a review of the main mechanisms controlling incremental oil recovery by low salinity water in sandstones was presented by Jackson et al. (2016). Their discussion was focused on mineral-surface-scale mechanisms. The authors believe that multiion exchange (MIE), local increase in pH, double layer expansion (DLE) are the main controllers. The latter mechanisms are common in changing zeta potential at mineral surfaces either through changing mineral surface charge or changing the double layer thickness. Hence, they stressed on the need for relating the referred mechanisms to zeta potential and measuring the latter at reservoir conditions. Kilybay et al. (2017) presented a review article on the main mechanisms controlling incremental oil recovery by low salinity/engineered water in both sandstone and carbonate rocks. This book is mainly a comprehensive review of the work performed so far in the area of LSWIs/EWIs and their applications in both sandstones and carbonates. In the context of this book, we refer to the process of diluting the injected water as LSWI; however, hardening or softening of the injected water is referred to as EWI. LSWI has been applied for both sandstones and carbonates with more emphasis on sandstones, whereas EWI has mostly been conducted for carbonates.

In Chapter 2, we present the experimental work of low salinity/engineered water conducted on both sandstones and carbonates. Chapter 3 presents field-scale applications of LSWI/EWI on sandstone and carbonate rocks. In Chapter 4, the underlying proposed mechanisms for incremental oil recovery by LSWI/EWI are described for both sandstone and carbonate rocks. Chapter 5 describes and discusses the different proposed LSWI/EWI models for both sandstones and carbonates. Chapter 6 reviews the basics of geochemistry and its application in the LSWI/EWI field including geochemical and compositional modeling approaches. The different applications of LSWI/EWI are discussed in Chapter 7 including conformance control and the combination of LSWI/EWI along with each of surfactants, polymers, and carbon dioxide (CO_2). In Chapter 8, we compare the main controlling factors for the effect of LSWI/EWI on both sandstone and carbonate rocks. Finally, Chapter 9 includes the main conclusions and recommendations based on this vast literature review and our experiences.

REFERENCES

Ayirala, S., Ernesto, U., Matzakos, A., Chin, R., Doe, P., Hoek, P.V.D. 2010. A designer water process for offshore low salinity and polymer flooding applications. Paper SPE 129926, SPE Improved Oil Recovery Symposium, Tulsa, OK.

Dandona, A.K., Morse, R.A., 1972. The influence of gas saturation on waterflood performance-variations caused by changes in flooding rate. Paper SPE 4257, SPE Hobbs Regional Meeting, Hobbs, NM.

Dang, C.T.Q., Nghiem, L.X., Chen, Z., Nguyen, Q.P., Nguyen, Ngoc. T. B, 2013. State-of-the art low salinity waterflooding for enhanced oil recovery. Paper SPE 165903, SPE Asia Pacific Oil & Gas Conference and Exhibition, Jakarta, Indonesia.

Guerithault, R., and Economides, C.A.E., 2001. Single-well waterflood strategy for accelerating oil recovery. Paper SPE 71608, SPE Annual Technical Conference and Exhibition, New Orleans, LO.

Jackson, M.D., Vinogradov, J., Hamon, G., Chamerois, M., 2016. Evidence, mechanism, and improved understanding of controlled salinity water injection part 1: sandstones. Fuel J. 185 (2016), 772–793.

Kilybay, A., Ghosh, B., Thomas, N.C., 2017. A review on the progress of ion-engineered water flooding. J. Petrol. Eng. 2017. Article ID: 7171957.

Lake, L.W., 1989. Enhanced Oil Recovery. Prentice Hall, Englewood Cliffs, NJ.

Sheng, J.J., 2014. Critical review of low-salinity waterflooding. J. Petrol. Sci. Eng. 120 (2014), 126–224.

Yousef, A.A., Ayirala, S.C., 2014. Optimization study of a novel water-ionic technology for smart-waterflooding application in carbonate reservoirs. Oil Gas Facilit. 3 (5), 72–82.

CHAPTER TWO

Experimental Works on Low Salinity/Engineered Water Injection

Contents

This chapter includes the LSWI/EWI effect on both sandstone and carbonate rocks at laboratory-scale.

2.1 LABORATORY APPLICATIONS OF LSWI/EWI ON SANDSTONE ROCK

For sandstone rocks, extensive laboratory studies have been conducted after producing 15% additional oil from Kansas field cores, when brine was used as injection fluid as opposed to fresh water (Smith, 1942). In the latter study, the decrease in oil recovery with injection of fresh water was due to clay swelling. Hence, Hughes and Pfister (1947) focused on fluid physical and chemical characteristics to prevent clay swelling. Reiter (1961) evaluated and compared the additional oil produced from Nacatoch sandstone oil-wet cores using both high-salinity water (Nacatoch connate water) and low salinity water (one-quarter of salinity of Nacatoch water). Results showed that low salinity water recovered 21.3% greater ultimate oil recovery than the more saline floods due to clay hydration effect. Bernard (1967) investigated relative effectiveness of fresh and salt waters on oil recovery from both synthetic and natural

Low Salinity and Engineered Water Injection for Sandstone and Carbonate Reservoirs.
DOI: http://dx.doi.org/10.1016/B978-0-12-813604-1.00002-X

cores, containing clays. Results showed no effect of saline water 15% to 1% NaCl on oil recovery; however, when the NaCl concentration was reduced from 1% to 0.1%, both oil recovery and pressure drop across the cores increased. Hence, the incremental oil recovery was related to the presence of clay. On the contrary, Al-Mumen (1990) reported an increase in oil recovery from Berea sandstone cores with increasing salinity up to a certain level after which recovery did not increase significantly.

Most of the researchers observed a certain salinity level needed to achieve incremental oil recovery by LSWI in sandstones. Several coreflood experiments were conducted by Zhang et al. (2007a) to investigate the effect of low salinity brine on improving the oil recovery of Berea sandstone cores oil for both secondary and tertiary modes. Results were promising for the salinity level of 1500 ppm NaCl; however, the use of 8000 ppm NaCl had no effect on oil recovery, as the reduction in the brine salinity level seemed insufficient. The applicability of low salinity waterflooding for Alaska North Slope (ANS) was evaluated by Patil et al. (2008) using coreflooding experiments in the secondary mode. They observed a reduction in residual oil saturation from 46% to 38% with decreasing salinity from 22,000 ppm to 5500 ppm. Moreover, oil recovery increased from 40% to 68% upon reducing water salinity from 22,000 ppm to ultralow salinity lake water (50–60 ppm). Webb et al. (2005a) observed low salinity effect on oil recovery at salinities in the range of 2000–3000 ppm. The previous studies showed that there is a salinity level below which low salinity effect on oil recovery is pronounced.

Both connate and invading brines have major effect on wettability and oil recovery at reservoir temperature, as was reported by Tang and Morrow (1997). Decreasing salinity of both connate and invading brines or either one causes increasing oil recovery, especially at high temperature, which increases the water wetness of the rocks resulting in high recovery. Brine salinity and temperature play a major role in enhancing oil recovery as was indicated by Agbalaka et al. (2009) through coreflooding experiments on Berea sandstone and shale sandstone core plugs at both low to high temperatures and using low salinity (<2 wt% NaCl) and high-salinity (4 wt% NaCl) brines. Additional oil recovery was obtained in both secondary and tertiary modes, when reducing the brine salinity from 4 to 1 wt%. The recovery increased even further at high temperatures compared to low temperatures.

Loahardjo et al. (2007) observed that sandstone reservoir cores responded better to low salinity water than outcrop cores. They concluded that low salinity waterflood is highly specific to crude oil–brine–rock

(COBR) interactions and cannot be predicted. This conclusion was based on spontaneous imbibition tests, which showed that the secondary recovery was enhanced by 16% OOIP for 10 × diluted seawater and 29% for 100 × dilution, whereas for the tertiary mode this was not always the case. The incremental oil recovery from sandstone rocks was in the range of 5–20% of OOIP as reported by most studies (Lager et al., 2007; Lager et al., 2008; Webb et al., 2005a; Webb et al., 2008).

Several researchers confirmed the low salinity water injection effect on oil recovery from sandstone rocks though a total of 214 coreflooding experiments in the secondary mode and 188 experiments in the tertiary mode (Aladasani et al., 2012). Gamage and Thyne (2011) showed the advantages of using low salinity water in the secondary mode of injection as opposed to the tertiary mode by producing 6–22% more oil through two-phase coreflooding experiments on Berea and field sandstone cores. Oil recovery due to low salinity water injection was coincident with decrease in pressure drop, increase in pH, and fine migration. Fjelde et al. (2012) conducted coreflooding experiments on sandstone reservoir core plugs from the North Sea to investigate the effect of low salinity water on oil recovery. They found that the high-salinity injected water resulted in a piston-like displacement, while the oil was recovered over a longer period using the low salinity water. This indicated that the rock was water-wet using high-salinity floods and converted to less water-wet using low salinity floods. They concluded that the direction of wettability alteration can be best explained by the ion exchange on the clay surface and, hence, analysis of LSWI/EWI effect should be performed on a case-by-case basis depending on interactions between formation brine, injected brine, oil components, and rock type.

Suijkerbuijk et al. (2012) studied the impact of injected water, formation water and crude oil on wettability alteration by LSWI/EWI. The study includes spontaneous imbibition experiments on both Berea and reservoir sandstone core plugs. They concluded that increasing the Ca^{2+} concentration in the injected water turns the system more oil-wet. Moreover, injection of NaCl in high-salinity and low salinity improves oil recovery, when the formation water has low divalent cations concentration and high divalent cations concentration, respectively. In addition, they found that the crude oil dictates the wettability state of the rock using LSWI/EWI and most oil-wet rocks resulted in pronounced incremental oil recovery. In a later work, Suijkerbuijk et al. (2014) evaluated the applicability of low salinity waterflooding on West Salym Field in

Russia through their spontaneous imbibition tests and corefloods on sandstone cores. The results showed a positive effect of low salinity waterflooding on oil recovery, which was justified by wettability alteration towards a more water-wet state by changing relative permeabilities and reducing residual oil saturation. Moreover, upscaling of coreflood results to field scale showed incremental oil recovery of 4% of OOIP in the secondary mode compared to 1.7% in the tertiary mode.

The previous studies show clearly the pronounced effect of low salinity on oil recovery from sandstones at laboratory-scale. In the next section, the effect of LSWI/EWI on carbonates at a laboratory-scale is discussed thoroughly.

2.2 LABORATORY APPLICATIONS OF LSWI/EWI ON CARBONATE ROCK

The effect of low salinity/engineered water injections on carbonates is not well covered compared to sandstones due to previous thoughts of relating wettability alteration by low salinity water to the presence of clay, which is not the case in carbonate rocks. Nevertheless, the effect of low salinity/engineered water injections on oil recovery from carbonate rocks was investigated at laboratory-scale using both spontaneous imbibition and coreflooding studies, and to a limited extent at field-scale.

2.2.1 Spontaneous imbibition tests

For spontaneous imbibition experiments, Hognesen et al. (2005) concluded from their experiments on reservoir limestone cores, outcrop chalk cores, seawater, and formation water that increasing sulfate ion concentration at high temperature leads to increasing oil recovery because of the role of sulfate ion as a wettability-modifying agent for carbonate rocks from mixed-wet to water-wet. Webb et al. (2005b) investigated the effect of sulfate on oil recovery from North Sea carbonate core samples through spontaneous imbibition experiments. They found that seawater has the ability to alter wettability of the carbonate system to a more water-wet state compared to sulfate free water. Moreover, Webb et al. (2005b) proved the effect of seawater on Valhall

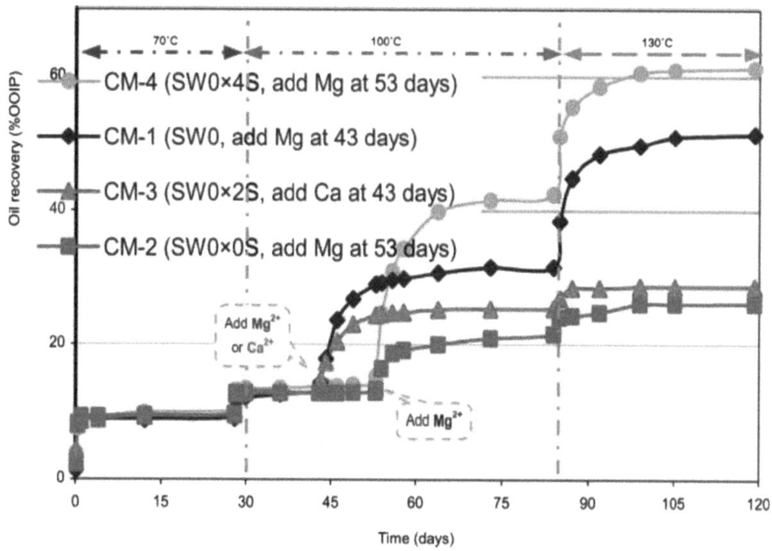

Figure 2.1 Wettability alteration proposed mechanisms (Zhang et al., 2007b).

chalk cores from the change in the capillary pressure curves with seawater injection compared to formation water. It was noticed that 40% of OOIP was recovered from chalk cores spontaneously using seawater compared to formation water and the recovery even increased from 40% to 60% with forced seawater injection.

Zhang et al. (2007b) studied wettability alteration of North Sea chalk reservoirs in Ekofisk field. The study was performed using an oil with AN of 2.07 mgKOH/g and NaCl brine with different concentrations of sulfate. Later, the effect of adding calcium or magnesium ions was investigated at various temperatures. As seen in Fig. 2.1, they concluded that wettability alteration occurs if the imbibing water contains either Ca^{2+} and SO_4^{2-} or Mg^{2+} and SO_4^{2-}.

Both spontaneous and forced displacement tests on chalk using a crude oil of AN = 1.9 mgKOH/g were conducted by Strand et al. (2008a). This work has verified the improvement in water wetness of carbonate rocks using seawater at high temperature, which leads to high oil recovery (Fig. 2.2).

Also, preliminary experimental studies for the chemical mechanism of wettability alteration in fractured limestone after seawater injection was performed by Strand et al. (2008b). The study showed a 15% increase in oil recovery from limestone cores, when seawater was injected compared to

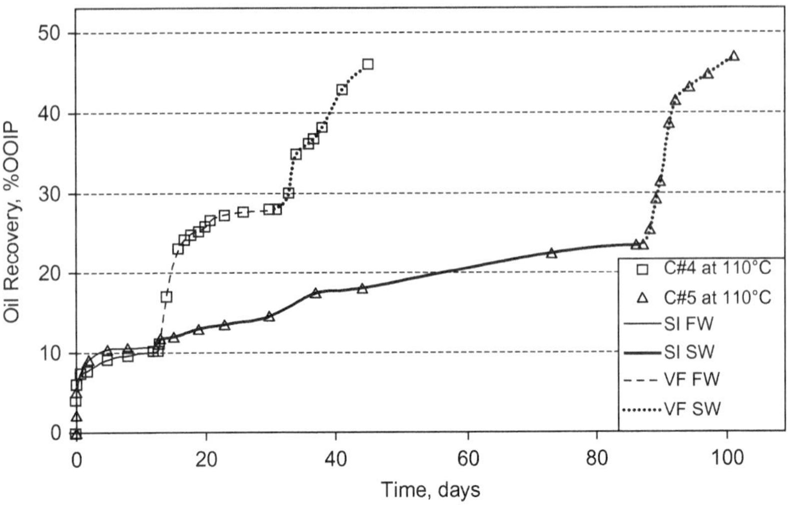

Figure 2.2 Effect of seawater injection on oil recovery from chalk cores compared to formation water both spontaneously and using forced displacement (Strand et al., 2008a).

seawater free of sulfate. Seawater was the lowest salinity injected brine compared to other higher salinity brines used; however, no lower salinity brine than seawater was used. This study also confirmed using chromatographic wettability tests that the effect of wettability alteration and sulfate adsorption is about the same for both chalk and limestone rocks. Increased oil recovery using low salinity water injection in limestone formations was noticed by Fjelde (2008). The results showed similar oil recovery by spontaneous imbibition of low salinity water compared to seawater experiments.

2.2.2 Coreflooding experiments

For coreflooding experiments, Bagci et al. (2001) reported high oil recovery of 35.5% OOIP by using 2 wt% KCl on limestone cores and high pH effluent brine due to ions exchange reactions with the clay present in the rock. They considered wettability alteration as the reason behind recovering more oil without further explanation. Yousef et al. (2011) investigated the applicability of low salinity water injection (smart waterflood) on carbonate rocks for improving oil recovery by using seawater and different dilutions of seawater. Results of coreflooding experiments showed increased oil recovery with stepwise dilution of seawater up to an 18% incremental oil recovery due to tertiary water

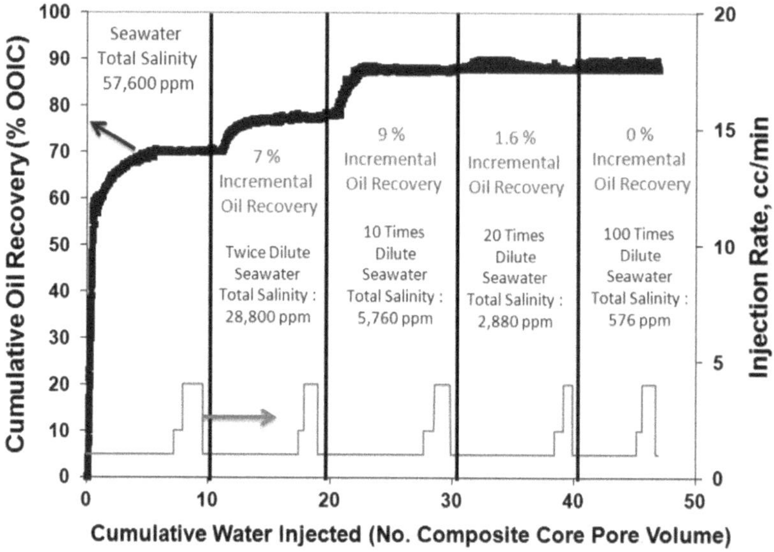

Figure 2.3 Oil recovery curve for the first coreflood of Yousef et al. (2011).

injection as seen in Fig. 2.3. About the same results were obtained by a second coreflooding experiment as a confirmation.

Later, researchers started turning the salinity of the injected water (engineered water injection) and checking the effect on oil recovery. Gupta et al. (2011) performed coreflooding experiments on both dolomite cores from West Texas and limestone cores from the Middle East. Experiments showed incremental 5–9% OOIP recovery from both dolomite and limestone cores because of adding sulfate ions. For limestone cores, 7–9% OOIP was obtained due to reducing hardness of the injected water, but not total dissolved solids. Another interesting finding is 15% and 20% OOIP by using borate (BO_3^{3-}) and phosphate (PO_4^{3-}) as modified ions, respectively. The study was performed in the tertiary mode after injecting formation water as base brine. They concluded that softening of injected brine contributes to dissolution and hardening contributes to changing surface charge, both of which are needed for the wettability alteration mechanism in carbonate rocks. Moreover, an increase in pH was observed through their corefloods on carbonates using modified water with PO_4^{3-}, BO_3^{3-}, and NaOH salts.

Zahid et al. (2012) conducted coreflooding experiments on both reservoir and outcrop chalk carbonate plugs to investigate the effect of LSWI through injection of seawater followed by different dilutions in succession. The results showed that there was no additional recovery from the outcrop chalk plugs at both high and low temperatures. Nevertheless, an additional recovery was obtained from reservoir carbonate plugs only at high temperatures combined with an increase in the pressure drop as a sign of either dissolution or fine migration. Calculations showed that the critical capillary number was exceeded in this experimental work, which is obvious from the incremental oil recovery obtained by changing the rate of injection. The capillary number is the ratio of viscous-to-capillary forces and is defined as (Abrams, 1975):

$$N_c = \frac{u\mu_w}{\sigma k_{rw}} = \frac{k\Delta P}{\sigma L} \tag{2.1}$$

where u is Darcy's injection velocity, μ_w is water viscosity, σ is oil-water interfacial tension, k_{rw} is water relative permeability, k is absolute permeability, and $\frac{\Delta P}{L}$ is the applied pressure gradient. The critical capillary number ($N_c^\star$) is the capillary number value at which the non-wetting phase saturation states to decrease (Donaldson et al., 1989). This excess of critical capillary number caused fine migration and loss in coherence of the core plugs, which led to an increase in pressure drop.

Other researchers claim no significant contribution for calcium ion on incremental oil recovery by LSWI/EWI. Chandrasekhar and Mohanty (2013) investigated the brines that improve oil recovery and the associated mechanisms through contact angle measurements, spontaneous imbibition, coreflood, and ion analysis on limestone reservoir rocks at high temperature of 248°F. The results showed that modified seawater containing Mg^{2+} and $SO_4{}^{2-}$ and diluted seawater alters the rock wettability to a more water-wet state; however, seawater with only Ca^{2+} was unsuccessful in changing wettability state. Moreover, the residual oil saturation to modified brine injection was about 20%. In addition, multi-ion exchange and mineral dissolution are responsible for desorption of organic acid groups and hence wettability alteration.

On the other hand, some researchers found an optimum concentration for potentially determining ions (Ca^{2+}, Mg^{2+}, and $SO_4{}^{2-}$). Al-Attar et al. (2013) evaluated the effect of applying low salinity/engineered water injections on carbonate cores from Bu Hasa field in Abu Dhabi using seawater and two field injection waters. Corefloods, interfacial tension (IFT), pH, and contact angle measurements were used at ambient conditions in the assessment, upon which they found an improvement in oil recovery from 63% to 84.5% of OOIP when diluting field water from

197,362 ppm to 5000 ppm. Moreover, they concluded that shifting the rock wettability from water-wet towards intermediate-wet is responsible for the incremental oil recovery obtained. In addition, they reported that pH and IFT could not justify the incremental oil recovery by low salinity water flooding. In addition, they found that increasing sulfate concentration up to a certain level has a positive effect on incremental oil recovery, whereas calcium concentration has a negative effect. Awolayo et al. (2014) investigated the impact of sulfate ion on incremental oil recovery by smart waterflood from carbonates. They concluded from their corefloods, contact angle measurements, zeta potential tests, and ionic analysis that the higher concentration of sulfate ion leads to higher oil recovery up to a certain limit. They suggested that smart water with four times sulfate concentrations might be the optimum sulfate concentration.

After discussing the experimental work of LSWI/EWI on both sandstones and carbonates, the field applications of this technology are discussed in the next chapter.

REFERENCES

Abrams, A., 1975. The influence of fluid viscosity, interfacial tension, and flow velocity on residual oil saturation left by waterflood. Soc. Petrol. Eng. J. 15 (5), 437–447.

Agbalaka, C.C., Dandekar, A.Y., Patil, S.L., Khataniar, S., Hemsath, J.R., 2009. Coreflooding studies to evaluate the impact of salinity and wettability on oil recovery efficiency. Transport Porous Med. 76 (1), 77–94.

Aladasani, A., Bai, B., Wu, U., 2012. Investigating low-salinity waterflooding recovery mechanisms in sandstone reservoirs. Symposium on SPE Improved Oil Recovery, Tulsa, Oklahoma, USA, Paper SPE 152997.

Al-Attar, H.H., Mahmoud, M.Y., Zekri, A.Y., Almehaideb, R.A., Ghannam, M.T., 2013. Low salinity flooding in a selected carbonate reservoir: Experimental approach. EAGE Annual Conference & Exhibition, London, United Kingdom, Paper SPE 164788.

Al-Mumen, A.A., 1990. The effect of injected water salinity on oil recovery. Master of Science Thesis, King Fahad University of Petroleum and Minerals, Dhahran, Saudi Arabia.

Awolayo, A., Sarma, H., AlSumaiti, A.M., 2014. A laboratory study of ionic effect of smart water for enhancing oil recovery in carbonate reservoirs. SPE EOR Conference at Oil and Gas West Asia, Muscat, Oman, Paper SPE 169662.

Bagci, S., Kok, M.V., Turksoy, U., 2001. Effect of brine composition on oil recovery by waterflooding. J. Petrol. Sci. Technol. 19 (3–4), 359–372.

Bernard, G.G., 1967. Effect of floodwater salinity on recovery of oil from cores containing clays. SPE California Regional Meeting, Los Angeles, California, USA, Paper SPE 1725.

Chandrasekhar, S., Mohanty, K.K., 2013. Wettability alteration with brine composition in high temperature carbonate reservoirs. SPE Annual Technical Conference and Exhibition, New Orleans, Louisiana, USA, Paper SPE 166280.

Donaldson, E.C., Chilingarian, G.V., Yen, T.F., 1989. Enhanced Oil Recovery II, Processes and Operations. Elsevier Science Publishers, Amsterdam, The Netherlands.

Fjelde, I., 2008. Low salinity water flooding experimental experience and challenges. Force RP Work Shop: Low salinity water flooding, the importance of salt content in injection water, Stavanger, Norway.
Fjelde, I., Asen, S.M., Omekeh, A., 2012. Low salinity water flooding experiments and interpretation by simulations. SPE Improved Oil Recovery Symposium, Tulsa, Oklahoma, USA, Paper SPE 154142.
Gamage, P., Thyne, G., 2011. Comparison of oil recovery by low salinity waterflooding in secondary and tertiary recovery modes. SPE Annual Technical Conference and Exhibition, Denver, Colorado, USA, Paper SPE 147375.
Gupta, R., Smith, G.G., Hu, L., Willingham, T., Cascio, M.L., Shyeh, J.J., et al., 2011. Enhanced waterflood for Middle East carbonates cores – Impact of injection water composition. SPE Middle East Oil and Gas Show and Conference, Manama, Bahrain, Paper SPE 142668.
Hognesen, E.J., Strand, S., Austad, T., 2005. Waterflooding of preferential oil-wet carbonates: Oil recovery related to reservoir temperature and brine composition. SPE EUROPEC/EAGE Annual Conference, Madrid, Spain, Paper SPE 94166.
Hughes, R.V., Pfister, R.J., 1947. Advantages of brines in secondary recovery of petroleum by waterflooding. Trans. AIME. 170 (1), 187–201.
Lager, A., Webb, K.J., Black, C.J.J., 2007. Impact of brine chemistry on oil recovery. 14th European Symposium on IOR, Cairo, Egypt.
Lager, A.K., Webb, K.J., Collins, I.R., Richmond, D.M., 2008. LoSal™ enhanced oil recovery: Evidence of enhanced oil recovery at the reservoir scale. SPE Symposium on Improved Oil Recovery, Tulsa, Oklahoma, USA, Paper SPE 113976.
Loahardjo, N., Xie, X., Yin, P., Morrow, N.R., 2007. Low salinity waterflooding of a reservoir rock. International Symposium of the Society of Core Analysts, Calgary, Alberta, Canada, Paper SCA2007–29.
Patil, S., Dandekar, A.Y., Patil, S.L., Khataniar, S., 2008. Low salinity brine injection for EOR on Alaska North Slope (ANS). International Petroleum Technology Conference, Kuala Lumpur, Malaysia, Paper SPE 12004.
Reiter, Pl. K., 1961. A water-sensitive sandstone flood using low salinity water. Master of Science Thesis, University of Oklahoma, USA.
Smith, K.W., 1942. Brines as flooding liquids. Seventh Annual Technical Meeting, Mineral Industries Experiment Station, Pennsylvania State College.
Strand, S., Puntervold, T., Austad, T., 2008a. Effect of temperature on enhanced oil recovery from mixed wet chalk cores by spontaneous imbibition and forced displacement using seawater. Energ. Fuel. 22 (5), 3222–3225.
Strand, S., Austad, T., Puntervold, T., Hognesen, E.J., Olsen, M., Barstad, S.M.F., 2008b. Smart water for oil recovery from fractured limestone: A preliminary study. Energ. Fuel. 22 (5), 3126–3133.
Suijkerbuijk, B.M.J.M., Hofman, J.P., Ligthelm, D.J., Romanuka, J., Brussee, N., van der Linde, H.A., et al., 2012. Fundamental investigations into wettability and low salinity flooding by parameter isolation. SPE Improved Oil Recovery Symposium, Tulsa, Oklahoma, USA, Paper SPE 154204.
Suijkerbuijk, B.M.J.M., Sorop, T.G., Parker, A.R., Masalmeh, S.K., Chmuzh, I.V., Karpan, V.M., et al., 2014. Low salinity waterflooding at West Salym: Laboratory experiments and field forecasts. SPE EOR Conference at Oil and Gas West Asia, Muscat, Oman, Paper SPE 169691.
Tang, G.Q., Morrow, N.R., 1997. Salinity temperature, oil composition and oil recovery by waterflooding. SPE Reserv. Eng. 12 (4), 269–276.
Webb, K.J., Black, C.J.J., Edmonds, I.J., 2005a. Low salinity oil recovery-the role of reservoir condition core floods. 13th European Symposium on Improved Oil Recovery, Budapest, Hungary.

Webb, K.J., Black, C.J.J., Tjetland, G., 2005b. A laboratory study investigating methods for improving oil recovery in carbonates. SPE International Petroleum Technology Conference, Doha, Qatar, Paper SPE 10506.

Webb, K., Lager, A., Black, C., 2008. Comparison of high/low salinity water/oil relative permeability. International Symposium of the Society of Core Analysts, Abu Dhabi, UAE, SCA2008–39.

Yousef, A.A., Al-Saleh, S., Al-Kaabi, A., Al-Jawfi, M., 2011. Laboratory investigation of the impact of injection-water salinity and ionic content on oil recovery from carbonate reservoirs. SPE Reserv. Eval. Eng. 14 (5), 578–593.

Zahid, A., Shapiro, A., Skauge, A., 2012. Experimental studies of low salinity water flooding in carbonate reservoirs: A new promising approach. SPE EOR Conference at Oil and Gas West Asia, Muscat, Oman, Paper SPE 155625.

Zhang, P., Tweheyo, M.T., Austad, T., 2007b. Wettability alteration and improved oil recovery by spontaneous imbibition of seawater into chalk: Impact of the potential determining ions Ca^{2+}, Mg^{2+}, and SO_4^{2-}. Colloids Surf. A Physicochem. Eng. Asp. 301 (1–3), 199–208.

Zhang, Y., Xie, X., Morrow, N.R., 2007a. Waterflood performance by injection of brine with different salinity for reservoir cores. SPE Annual Technical Conference and Exhibition, Anaheim, California, USA, Paper SPE 109849.

CHAPTER THREE

Field Applications of Low Salinity/Engineered Water Injection

Contents

This chapter describes the field applications of the low salinity/engineered water injection (LSWI/EWI) technique for mainly sandstone rocks as there is limited application in carbonates so far.

3.1 FIELD APPLICATIONS OF LSWI/EWI ON SANDSTONE ROCKS

Field-scale studies were conducted as well for investigating the effect of LSWI on oil recovery from sandstone rocks. Webb et al. (2004) confirmed the laboratory results at field-scale using a modified log–inject–log on a giant clastic reservoir, measuring the waterflood residual oil saturation (S_{orw}) after both high and LSWIs. Laboratory studies showed that there is insignificant change in oil saturation for injected brines of salinity over 35,000 ppm. This was consistent with the result of Webb et al. (2004) as there was no significant difference between high and intermediate salinities, and both of them were used for defining the base remaining oil for LSWI. Fig. 3.1 shows results of LSWI, as the change in remaining oil saturation is obvious between high-salinity and LSWIs. Remaining oil saturation out of this study ranged from 30% to 50% which matched previous laboratory studies. Possible modes of LSWI could be applied based on this optimistic result, including low salinity

DOI: http://dx.doi.org/10.1016/B978-0-12-813604-1.00003-1

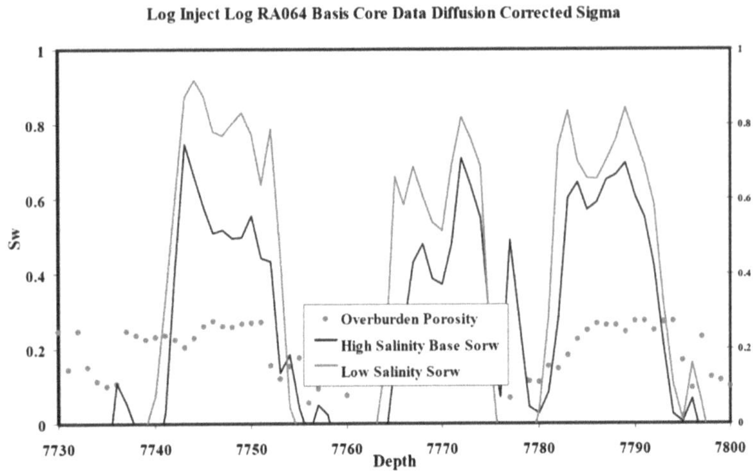

Figure 3.1 Remaining oil saturation post low salinity waterflood (Webb et al., 2004).

alternating regular salinity, low salinity slug, hot low salinity water, low salinity water-alternating-gas (LSWAG), and others (Webb et al., 2004).

Single-well chemical tracer tests (SWCTTs) were conducted in Alaska as was reported by McGuire et al. (2005). Results showed an increase in oil recovery between 8% and 19% due to LSWI. The authors reported a limit of 5000 ppm salinity below which LSWI effect is pronounced. Upper critical salinity limit was observed in several papers for having a prominent salinity effect on oil recovery. Another SWCTT was conducted to measure the residual oil before and after injection of low salinity brine in the tertiary mode as was reported by Seccombe et al. (2008). A 10% decrease in residual oil saturation was observed because of LSWI (2600 ppm) in two producers. The first comprehensive inter-well application was reported by Seccombe et al. (2010) involving an injector and a producer 1040 ft apart. The results were in agreement with the corefloods, single-well tracer tests, and incremental oil recovery versus clay content relationship.

Moreover, Vledder et al. (2010) described a field-scale proof of wettability alteration using LSWI in mixed to oil-wet sandstone reservoirs in the Omar Field in Syria. The reservoir has a light oil of 0.3 cP viscosity, formation water salinity of 90,000 ppm, and bivalent cations of 5000 ppm. The LSWI was conducted in the secondary mode using the river water of 500 ppm salinity and less than 100 ppm bivalent cations. The change in wettability was described using the extended Buckley–Leverett theory

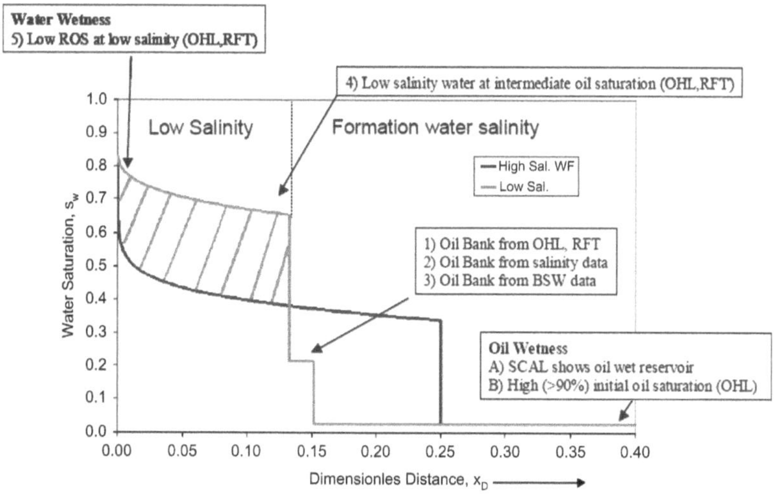

Figure 3.2 Saturation profile for both high and low salinity water flooding based on Buckley–Leverett theory (Vledder et al., 2010).

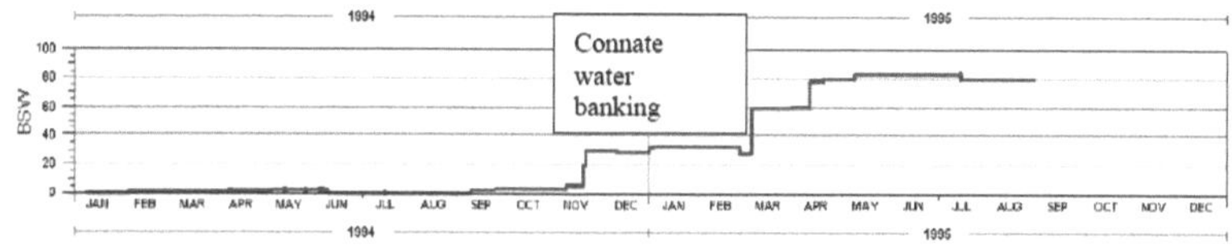

Figure 3.3 Water cut development with connate water banking between November 1994 and March 1995 (Vledder et al., 2010).

(Pope, 1980) with observations of delaying water breakthrough, oil banking ahead of the low salinity shock front due to accumulation of the desorbed oil, and reduction of remaining oil saturation behind the shock front (Fig. 3.2). The change in wettability was observed from the dual steps in water cut development (Fig. 3.3) and supported by spontaneous imbibition experiments on core plugs and single well log–inject–log tests. The results showed that about 10%–15% of OOIP can be recovered using LSWI in mixed-to-oil wet sandstone reservoirs.

Seven LSWI field pilots were performed in the Pervomaiskoye field in Russia, which resulted in incremental oil recovery of 5%–9% OOIP (Akhmetgareev and Khisamov, 2015). The authors investigated the mechanism behind this incremental oil recovery through conducting several double displacement tests using sandstone cores of different clay content. They reported that wettability alteration of rock surface was the main mechanism for cores with low clay content as twice-decreased residual oil saturation was observed. On the other hand, formation damage by clay migration was the reason for incremental oil recovery by LSWI on cores with high clay content, as two-time reduced water relative permeability was noticed. Moreover, the authors performed numerical modeling of the past-mentioned mechanisms on a field-scale using a fine-assisted model.

Unsuccessful field pilots of LSWI are also reported in the literature. Skrettingland et al. (2011) reported low or no effect of LSWI on oil recovery at the Snorre field in the North Sea. This finding was concluded from the SWCTT conducted in the tertiary injection mode and reasoned by the prevailing weakly water-wet conditions in this field. Callegaro et al. (2014) conducted a SWCTT in a complex clastic reservoir in West Africa. The test was designed to measure residual oil saturation after three consecutive injection cycles including seawater, low salinity water, and surfactant. The field results of low salinity flooding were not promising compared to laboratory observations. They explained this discrepancy due to the mineralogy in the tested formation that contains lower clay content and to the low residual oil saturation to water where less action is available for IOR treatment by LSWI.

3.2 FIELD APPLICATIONS OF LSWI/EWI ON CARBONATE ROCKS

The first-ever LSWI application in carbonate reservoirs was reported by Yousef et al. (2012). Two SWCTTs were applied in an Upper Jurassic carbonate reservoir using a diluted version of Qurayyah seawater. The tests resulted in about 7 saturation units reduction in the residual oil beyond conventional seawater injection. Two wells (A and B) were selected for the tests based on certain screening criteria. The objective of the test on well A was to confirm the effect of LSWI in the tertiary mode where three slugs were injected: two seawater slugs to assure achieving residual oil saturation conditions, followed by the smart water slug (10 times diluted). After each slug, three tracers were injected to determine the reduction in residual oil saturation. The SWCTT on well B was conducted to determine the effect of various dilutions of smart water on residual oil saturation, where three slugs were injected: a seawater slug, followed by a twice-diluted smart water slug, and finally a 10

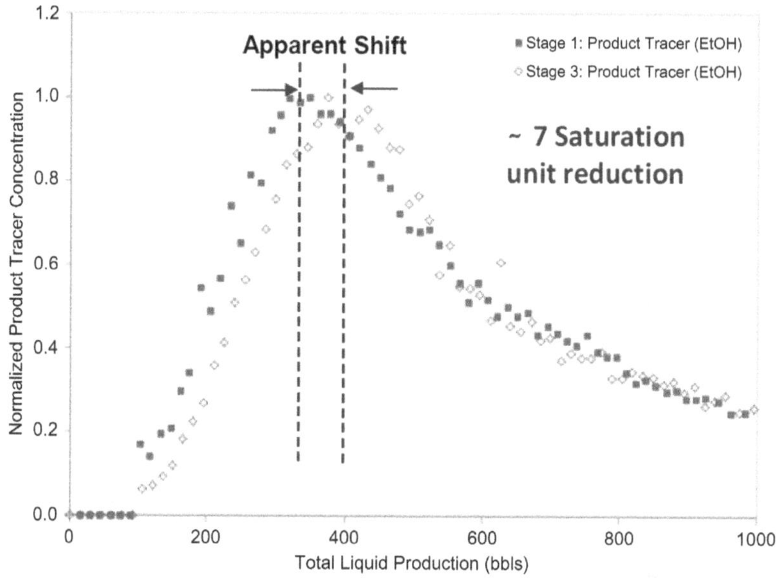

Figure 3.4 Comparison of field-measured tracer concentration profiles between Stage 1 (field seawater) and Stage 3 (SmartWater—10 times diluted) for Well A (Yousef et al., 2012).

times diluted smart water slug. Results obtained matched with their previous experimental work (Yousef et al., 2011), which is encouraging to plan a multi-well demonstration pilot. Fig. 3.4 shows an example of the results obtained for this field experience for well A.

Extensive research is still underway to understand the mechanism underlying incremental oil recovery by LSWI and hence expand successful field-scale applications. Hence, the mechanisms underlying the incremental oil recovery by LSWI/EWI from sandstone and carbonate rocks are discussed in the next chapter.

REFERENCES

Akhmetgareev, V., Khisamov, R., 2015. 40 years of low-salinity waterflooding in Pervomaiskoye Field, Russia: incremental oil. Paper SPE 174182, SPE European Formation Damage Conference and Exhibition, Budapest, Hungary.

Callegaro, C., Masserano, F., Bartosek, M., Buscaglia, R., Visintin, R., Hartvig, S.K., et al., 2014. Single well chemical tracer tests to assess low salinity water and surfactant EOR processes in West Africa. Paper SPE 17951, SPE International Petroleum Technology Conference, Kuala Lumpur, Malaysia.

McGuire, P.L., Chatham, J.R., Paskvan, F.K., Sommer, D.M., Carini, F.H., 2005. Low salinity oil recovery: an exciting new EOR opportunity for Alaska's North Slope. Paper SPE 93903, SPE Western Regional Meeting, Irvine, CA.

Pope, G.A., 1980. The application of fractional flow theory to enhanced oil recovery. SPEJ. 20 (3), 191–205.
Seccombe, J.C., Lager, A., Webb, K., Jerauld, G., Fueg, E., 2008. Improving waterflood recovery: LoSalTM EOR field evaluation. Paper SPE 113480, SPE Improved Oil Recovery Symposium, Tulsa, OK.
Seccombe, J., Lager, A., Jerauld, G., Jhaveri, B., Buikema, T., Bassler, S., et al., 2010. Demonstration of low-salinity EOR at interwell scale, Endicott Field, Alaska. Paper SPE 129692, SPE Improved Oil Recovery Symposium, Tulsa, OK.
Skrettingland, K., Holt, T., Tweheyo, M.T., Skjevark, I., 2011. Snorre low salinity water injection-coreflooding experiments and single well field pilot. SPE Reserv. Eval. Eng. 14 (2), 182–192.
Vledder, P., Fonseca, J.C., Wells, T., Gonzalez, I., Ligthelm, D., 2010. Low salinity water flooding: proof of wettability alteration on a field wide scale. Paper SPE 129564, SPE Improved Oil Recovery Symposium, Tulsa, OK.
Webb, K.J., Black, C.J.J., Al-Ajeel, H., 2004. Low salinity oil recovery-log-inject-log. Paper SPE 89379, SPE Symposium on Improved Oil Recovery, Tulsa, OK.
Yousef, A.A., Al-Saleh, S., Al-Kaabi, A., Al-Jawfi, M., 2011. Laboratory Investigation of the impact of injection-water salinity and ionic content on oil recovery from carbonate reservoirs. SPE Reserv. Eval. Eng. 14 (5), 578–593.
Yousef, A.A., Liu, J., Blanchard, G., Al-Saleh, S., Al-Zahrani, T., Al-Zahrani, R., et al., 2012. SmartWater flooding: industry's first field test in carbonate reservoirs. Paper SPE 159526, SPE Annual Technical Conference and Exhibition, San Antonio, TX.

CHAPTER FOUR

Mechanisms Behind LSWI/EWI Effect on Oil Recovery

Contents

In the literature, there is no consensus on a single underlining mechanism for incremental oil recovery caused by low salinity water injection in both sandstone and carbonate rocks. The different proposed mechanisms underlying incremental oil recovery by the low salinity/engineered water injection in both sandstones and carbonates are summarized in Fig. 4.1 and are discussed in this chapter.

4.1 PROPOSED MECHANISMS FOR LSWI/EWI ON SANDSTONE ROCKS

Different mechanisms behind low salinity/engineered water process have been proposed in light of enormous research studies; however, no single suggested mechanism has been accepted as the main mechanism for low salinity effect on sandstone rocks. This is due to the complex nature of the interaction between crude oil, brine, and rock, as well as due to a number of conflicting observations from one suggested mechanism to another. Previously suggested mechanisms include fines migration, pH increase, multiion exchange (MIE), salting-in, and wettability alteration

DOI: http://dx.doi.org/10.1016/B978-0-12-813604-1.00004-3

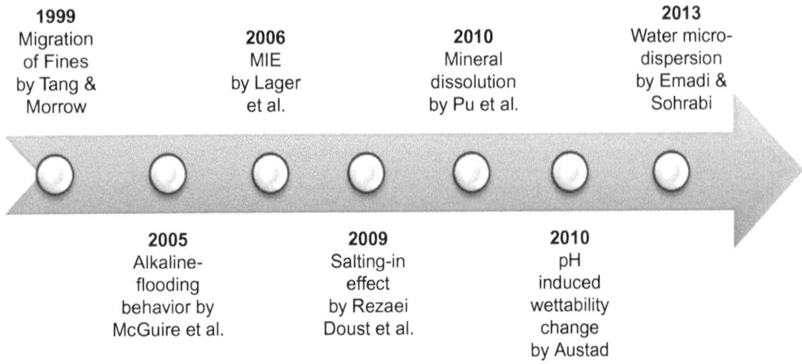

Figure 4.1 Timeline of the low salinity proposed mechanisms (Rotondi et al., 2014).

(Doust et al., 2009; Ligthelm et al., 2009). A description of each of the suggested mechanisms is included in this section.

4.1.1 Fine migration

Smith (1942) conducted several laboratory tests on sandstone cores using calcium chloride solution and fresh water. Oil recovery was higher for brine solution compared to fresh water and the reason was the swelling of clays in case of fresh water. Enhanced oil recovery by low salinity waterflooding on sandstone cores was performed by Martin (1959), who documented increasing oil recovery as a result of reducing the injection brine salinity and proposed swelling of clays and emulsification as reasons behind this increase. Nevertheless, Bernard (1967) found higher oil recovery using fresh water in sandstone compared to NaCl brine and this was accompanied by high pressure drop. Two scenarios were suggested: Swelling of clay leading to less space for oil and water and, hence, increasing oil recovery, and clay dispersion to fine particles blocking the originally established flow channels and forming new channels where oil recovery increases (Alotaibi and Nasr-El-Din, 2009).

Fine migration chemically initiated from clay occurs once the flowing fluid has an inadequate total cation concentration or contains inadequate percentage of divalent cations (Ca^{2+} and Mg^{2+}). Dispersion of clays is considered a complex phenomenon depending on the clay type and brine composition of both flowing and initially existing waters. Tang and Morrow (1999) proposed the fine migration mechanism, as the low salinity water could release clay fragments (fines), especially kaolinite detachments, rendering clay minerals more water-wet. Nevertheless, they highlighted the reason for the improvement in microscopic displacement efficiency is due to blocking of pore throats by the released clay particles and diverting water flow into nonswept pores rather than the wettability alteration effect. Doust et al. (2010) summarized that the aging temperature of 90°C is optimum for the low salinity water injection effect, regardless of the flooding temperature, and that clay content controls the low salinity effect on oil recovery; as an increase in the clay content results in increasing the additional oil recovery obtained by low salinity water injection. Another interesting finding is that adsorption of organic material is essential for a pronounced low salinity water effect in sandstone rocks.

Other researchers did not observe fine migrations in their studies, although low salinity effect was prominent (Lager et al., 2008). Fine migration is considered as an auxiliary mechanism for low salinity water injection, but not the main.

4.1.2 pH increase

Increasing pH of the system due to low salinity water injection was suggested as the reason for the incremental oil recovery obtained by McGuire et al. in 2005 by forming an alkaline flood. Lager et al. in 2006 disagreed by pointing out the need for generating in situ surfactant requires a crude oil with acid number (AN) >0.2 mgKOH/g; however, in some cases, the salinity effect was observed on oil samples with AN less than 0.05 mgKOH/g. Moreover, the increase in pH as noticed in many cases is not more than 1 pH unit, which makes the medium slightly basic and cannot justify the obtained increments in oil recovery. In addition, the decrease in interfacial tension (IFT) to ultralow values was not pronounced in the literature.

Austad et al. (2010) proposed a chemical mechanism for the effect of low salinity water injection. This mechanism suggests that thermodynamic chemical equilibrium exists initially at reservoir conditions of pH (5−6). The low pH environment enhances the adsorption of both acidic and basic components onto the clay surface. The chemical equilibrium is disturbed the moment low salinity water is injected; this causes interaction between brine and rock to compensate for the loss of cations, especially Ca^{2+}. This causes H^+ ions to get closer to compensate for the Ca^{2+} loss into the low saline water, which creates an increase in the pH close to the clay surface. Local increase in pH leads to reactions between the adsorbed basic and acidic material.

Clay minerals are unique by their cation exchange phenomenon. As a result of failure in the chemical structure, most of the clay minerals are permanently negatively charged and so they need cations for the balance to occur. This is usually satisfied by Ca^{2+} cations; as a result of low salinity water injection, exchanging of Ca^{2+} ions by H^+ ions at clay surface occurs according to Eq. (4.1), which results in a pH increase:

$$\left[\text{Clay}^-.......\text{Ca}^{2+}\right] + \text{H}_2\text{O} \leftrightarrow \left[\text{Clay}^-......\text{H}^+\right] + \text{OH}^- + \text{Ca}^{2+} \tag{4.1}$$

The reversibility of the pH can occur due to the injection of high-salinity water, which leads to exchanging H^+ ions by Ca^{2+} ions at clay surface and decreasing pH according to Eq. (4.2):

$$\left[\text{Clay}^-......\text{H}^+\right] + \text{Ca}^{2+} \leftrightarrow \left[\text{Clay}^-......\text{Ca}^{2+}\right] + \text{H}^+ \tag{4.2}$$

4.1.3 Multiion exchange

MIE mechanism, proposed by Lager et al. (2006), states that increase of water wetness due to low salinity water injection is the reason behind improving oil recovery by exchanging ions affecting the interactions between the clay minerals and surface-active components in the crude oil controlled by the presence of divalent ions such as Ca^{2+} and Mg^{2+}. An adsorption model was suggested, where Ca^{2+} acts like a bridge between the negatively charged clay surface and the negatively charged carboxylic material; this organic material is removed by cationic ion exchange at the surface.

4.1.4 Salting-in

Doust et al. (2009) proposed salting-in mechanism, which states that due to disturbing the thermodynamic equilibrium between the phases (water/oil/rock) by injecting low salinity water, the solubility of polar organic components in water changes. Salting-out and salting-in terminologies are well known in the chemical literature, as adding salt to the system (salting-out) leads to a decrease of organic material solubility in water; however, salting-in, which is decreasing the salinity of the system by salt removal from the water, leads to an increase in the solubility of these organic material. Organic materials are solvated in water by formation of a water structure using hydrogen bonds around the hydrophobic part.

Nevertheless, the presence of the inorganic material (Ca^{2+}, Mg^{2+}, and Na^{+}) leads to breaking this water structure and decreasing the solubility of these organic molecules. Thus, the concentration of the divalent ions has a much stronger effect on the solubility of the organic material in water. Decreasing the salinity of the system below a critical ionic strength can increase the solubility of the organic material in the aqueous phase, which is called the "salting-in" effect. Doust et al. (2009) verified this mechanism through a preliminary study as desorption of 4-tert-butyl benzoic acid in an aqueous suspension of kaolinite increases with the decreasing salinity of brine used.

4.1.5 Wettability alteration in sandstones

Extensive research work was performed for investigating the effect of low salinity water injection on changing the wettability of sandstone rocks. The conditions for the low salinity water injection effect to occur can be summarized from the work conducted by Tang and Morrow (1999) and Lager et al. (2007; 2008). Sandstone as a porous medium must contain clay and the type of clay mineral might play a role. Oil must have polar components and the injected water must contain divalent cations like Ca^{2+} and Mg^{2+}. The salinity of the injected water was usually between 1000 and 2000 ppm; effects were noticed even up to 5000 ppm. The concentration of Ca^{2+} vs Na^{+} plays a role and there is a little increase in pH of the effluent water. Fines migration that might occur as pressure increases across the core was observed. No temperature limitation was reported; most of the experiments were performed at temperatures below 100°C.

Wettability alteration is considered as the main phenomenon behind increasing oil recovery using low salinity water through various mechanisms including fine migration, pH increase leading to IFT reduction, MIE, and double layer expansion. The wettability alteration in sandstone rocks was related to the presence of clay minerals, oil composition, formation water with high concentration of divalent cations (Ca^{2+}, Mg^{2+}), and salinity level of water (1000–5000 ppm) (Tang and Morrow, 1997; Suijkerbuijk et al., 2012).

Wettability alteration mechanism and oil recovery were reported to be similar to those occurring during alkaline and surfactant waterflooding by McGuire et al. (2005). During low salinity water injection (1500 ppm) experiments into sandstone core, reactions occurred and pH increased up to 9, causing generation of surfactant, which lowers the IFT between oil and water and increases the water wettability leading to higher oil recovery. Zhang and Morrow (2006) suggested similar mechanism and noticed an increase in the effluent brine pH, which never exceeded 10. Clay-free sandstones provided evidence for the importance of clays for reaction with low salinity water. Moreover, similar reaction mechanism was suggested by Austad (2008), as with decreasing salinity below a critical value, the solubility of the organic materials in the aqueous phase occurs (salting-in phenomenon). As salinity of system decreases and thermodynamic equilibrium is disturbed, the cations desorb from the clay surface, breaking the bridge between the clay surface and the organic material and leading to desorption of these organic material from the clay surface. The release of these cations causes increase in the pH of the new equilibrium solution. This increase in the pH of the system helps in releasing more of the organic material and rendering the system more water-wet (Doust et al., 2009).

Lee et al. (2010) illustrated the wettability alteration mechanism in sandstones through double layer expansion, as there are two layers of ions surrounding the clay surface. The first layer is the "stern layer" which contains only cations (i.e., Na^+, Ca^{2+}, Mg^{2+}, etc.), while the second layer is called the "diffuse layer", which is far from the rock surface and where negative oil components might have the chance to enter this layer. The negative oil components have the potential to be attached to the divalents (i.e., Na^+, Ca^{2+}, Mg^{2+}, etc.) present in the stern layer, which turns the clay surface oil-wet. In the case of low salinity/engineered water injection, the double layer expands, the thickness of the double layer increases, the chance of oil components to get close to the stern layer and attach to the divalents decreases, and hence, the wettability changes more towards the water-wet condition (Fig. 4.2).

Nasralla and Nasr-El-Din (2011) investigated the effect of cation type and concentration in the injected water on oil recovery from Berea sandstone cores. Wettability alteration was reported to be the reason for oil recovery by changing the electrical charge for both oil/brine and rock/brine interfaces to highly negative, promoting repulsive electrical forces in the double layer and resulting in a more stable water film and water wetting state. In addition, they reported the dominance of cation type effect on oil recovery compared to the water salinity. Nasralla et al. (2011a) supported their previous findings of wettability alteration by cation exchange and reported the advantage of applying low salinity water injection in the secondary mode compared to the tertiary mode. Moreover, in another work Nasralla et al. (2011b) showed that contact

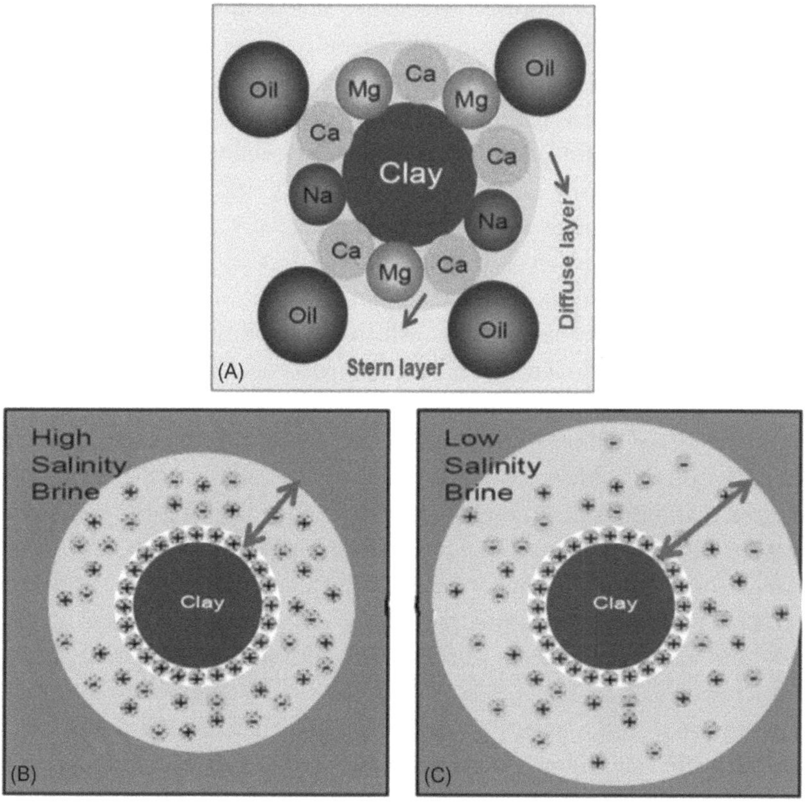

Figure 4.2 (A) Schematic of electric double layer and oil components adsorbed to the divalents through the double layer; (B) the thickness of double layer when high-salinity water is in contact with the clay surface; (C) the thickness of double layer when low salinity water is in contact with the clay surface (Lee et al., 2010).

angle increases with increasing temperature and pressure, and decreases with decreasing water salinity. In a later work, Nasralla and Nasr-El-Din (2012) investigated the expansion of the electric double layer as a primary mechanism by low salinity water injection through different corefloods on Berea sandstone cores. Results of this study showed that the double layer expansion mechanism is dominant in improving oil recovery by LSWI in the secondary mode. Nevertheless, no additional oil recovery was reported in the tertiary mode as they proposed that the discontinuity of oil phase mitigates expansion of the double layer. Moreover, reduction of pH of low salinity injected water reduces repulsion forces between oil and rock and results in a more unfavorable oil-wet system.

Rivet et al. (2010) reported wettability alteration to be the reason for the incremental oil recovery due to low salinity water injection from their corefloods on both Berea and reservoir cores. This mechanism was supported by observations of decreasing end-point water relative permeability and increasing end-point oil relative permeability. They reported a more stable water front case when the system is water-wet, which helps in delaying breakthrough and improving oil recovery. Moreover, although the residual oil saturation is lowest when rock is mixed-wet, water-wet case results in the most favorable mobility ratio. Mahani et al. (2014) investigated the mechanism by which wettability alteration occurs through their glass micromodel experiments on clay substrate. They observed the decrease in contact angle after injecting high-salinity water followed by low salinity water. They stressed on the importance of the kinetics of oil detachment when exposed to low salinity brine. Also, they reported that diffusion alone is not responsible for the long-time observed for wettability alteration. An electro-kinetic ion-transport can explain the delay in formation of cation-bridging, direct chemical bonds or acid/base effects. It is worth mentioning that the viscous force was absent in this study and the main present forces were buoyant and adhesion forces. Shehata and Nasr-El-Din (2015) conducted several zeta potential experiments to evaluate the effects of electrical surface charge and double layer expansion for common sandstone minerals. Results showed that the monovalent cations are more efficient in increasing the absolute values of zeta potential than divalent cations at 25°C. Moreover, with decreasing the salinity of the injected brine, the zeta potential becomes more negative.

Sohrabi et al. (2015) focused on fluid–fluid interactions as the cause of wettability alteration by LSWI. They reported that the injection of low salinity water rearranges the natural active surface component of crude oil, which coalesces in the form of micro-dispersions in the oil phase. These micro-dispersions desorb the previously adsorbed natural active surface component from the rock interface rendering the rock surface more water-wet. The authors supported their findings through direct flow visualization (micromodel), crude oil characterization through infrared spectroscopy, coreflood, and spontaneous imbibition experiments. They claimed that the proposed mechanism works for both sandstones and carbonates, and insensitive to the presence of clay.

Some of the researchers believe that the effect of low salinity water on incremental oil recovery should be interpreted at the nanoscale rather than at the microscale. Hassenkam et al. (2012) utilized atomic force microscopy (AFM) to justify the incremental oil recovery obtained by LSWI from sandstones by measuring and mapping of adhesion forces. In the experimental work, the authors used surfaces from real core samples, carboxylic acid end groups, and artificial seawater solutions with salinity of 36,000 and 1500 ppm. They observed a pronounced decrease in adhesion forces between the quartz grain surfaces and carboxylic acid when decreasing the salinity of brine solution from 36,000 ppm to 1500 ppm, which might be responsible for incremental oil recovery by LSWI. In a later work, Hassenkam et al. (2014) further highlighted the use of force mapping combined with core plug testing for gathering information about incremental oil recovery by LSWI. They observed a salinity threshold of 5000 to 8000 ppm for adhesion forces to decrease, which is similar to that of experimental core plugs and reservoir tests. Hence, nanometer scale observations complement and explain those of micrometer and even kilometer scales.

After discussing the low salinity waterflooding mechanisms in sandstones, the mechanisms in carbonates are discussed in the section below.

4.2 PROPOSED MECHANISMS FOR LSWI/EWI ON CARBONATE ROCKS

The mechanism underlying incremental oil recovery by LSWI/EWI in carbonates is less complicated compared to sandstones, as most of the researchers believe in wettability alteration.

4.2.1 Wettability alteration in carbonates

The thermodynamic equilibrium has long been established between the rock/oil/formation brine systems initially in the reservoir. Nevertheless, this equilibrium is not favorable in certain cases concerning wettability phenomenon, especially in carbonate rocks. Extensive research was performed by Austad and other coworkers (Standnes and Austad, 2000; Hognesen et al., 2005; Zhang et al., 2006; Puntervold et al., 2007), which opened the possibility of wettability alteration and enhancement of oil recovery from carbonate rocks by modifying the ionic composition in the injected water. Wettability alteration is the main and most acceptable mechanism for the incremental oil recovery achieved in carbonate rocks with the use of low salinity/engineered water injections. The wettability alteration phenomenon can occur due to change in the surface charge of the rock by desorption of organic material or dissolution.

For the case of changing the surface charge of the rock, several researchers related wettability alteration by low salinity water to sulfate adsorption on rock surface. Strand et al. (2003) investigated the sulfate concentration on wettability alteration with and without cationic surfactant solution using chalk, dolomite cores, and different crystals of carbonate rocks for spontaneous imbibition. They found that the effect of sulfate as a catalyst for imbibition rate in the presence of surfactants is more pronounced at high temperature and below a certain concentration of 1.0 g/L. The role of sulfate as a catalyst for improving imbibition rate comes from adherence on the rock surface. This makes this location partially negatively charged due to the presence of other positively charged metal ions. They also concluded that both sulfates and cation surfactants affect wettability alteration, turning carbonate rocks more water-wet at different degrees regarding carbonate rock type (calcite, dolomite, and magnesite) as shown in Fig. 4.3.

Moreover, Hognesen et al. (2005) conducted imbibition experiments on chalk and limestone cores for investigating the effect of cationic surfactants as well as sulfates as wettability modifiers using seawater. They found that with increasing the temperature (90–130°C), sulfate acts as a

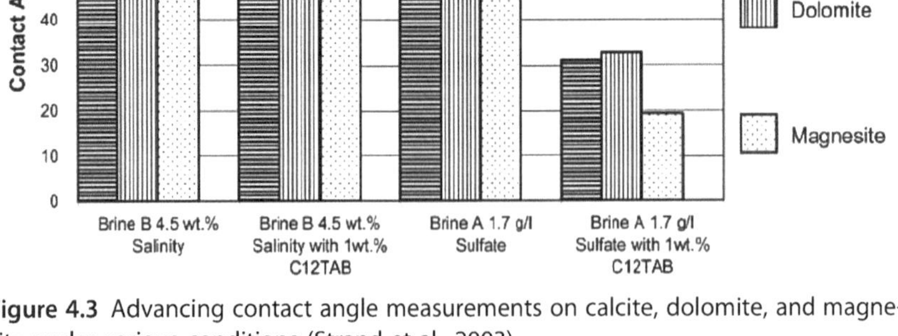

Figure 4.3 Advancing contact angle measurements on calcite, dolomite, and magnesite under various conditions (Strand et al., 2003).

catalyst causing an increase in oil recovery by imbibition in the presence of surfactants. The sulfate concentration used was 2.31 g/L, even higher than the sulfate concentration used by Strand et al. (2003), which resulted in a higher oil recovery due to the increase of sulfate adsorption on surface. The increase in sulfate affinity to carbonate surface with increasing temperature was verified to be the reason for such sulfate catalytic behavior at higher temperatures. The increase in the affinity of sulfate changes the rock charge locally from positive to negative and causes repulsion with carboxylic group, rendering the system water-wet. The use of cationic surfactants and sulfates is in lowering IFT besides wettability alteration. Therefore, increasing the temperature of the system not only decomposes the carboxylic group, but also increases the adsorption of sulfates on the rock surface, which promotes retention of water wetness. They stated that the benefit in using sulfate as a wettability modifier has limitations depending on initial brine salinity and temperature, as the concentration of Ca^{2+} in the connate brine should be known with certainty to avoid $CaSO_4$ precipitation.

Wettability alteration in carbonate rocks using LSWI/EWI can be achieved by injecting water containing SO_4^{2-} and either Ca^{2+} or Mg^{2+} or both of them in the presence of high temperature (>90°C). Both proposed mechanisms for wettability alteration in carbonate rocks are shown in Fig. 4.4. It was proposed that with increasing temperature, the affinity of sulfate to chalk rock surface increases and sulfate adsorption occurs. At the same time, Ca^{2+} adsorption increases as well as the initial positive

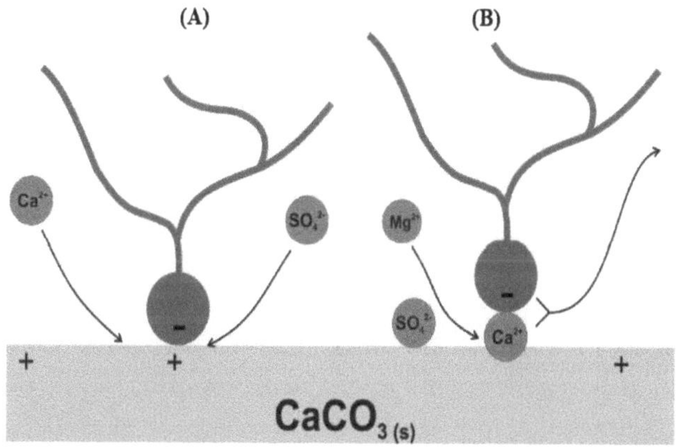

Figure 4.4 Proposed mechanisms for wettability alteration in carbonate rocks (Zhang et al., 2006).

charge of the rock decreases. Hence, more excess Ca^{2+} ions are present close to the surface, which reacts with the carboxylic material and releases some of them. Moreover, with increasing temperature, Mg^{2+} becomes more active, Ca^{+2} substitution by Mg^{2+} occurs, and sulfate becomes less active as it reacts with Mg^{2+}. Otherwise, $CaSO_4$ precipitation occurs and causes injection problem. This is shown in Fig. 4.4B, while Fig. 4.4A shows the effect of low temperature (less than 100°C), where Mg^{2+} becomes less active and both Ca^{2+} and $SO_4{}^{2-}$ becomes more active and $CaSO_4$ might occur (Zhang et al., 2006). Both wettability alteration mechanisms by Ca^{2+} and $SO_4{}^{2-}$ and by Mg^{2+} and $SO_4{}^{2-}$ at the chalk surface were previously shown in Fig. 2.1 (Chapter 2), which shows experimental results of the work performed by Zhang et al. (2007) confirming their proposed mechanism.

Previous studies indicated that wettability alteration process is related to the presence of Ca^{2+}, Mg^{2+}, and $SO_4{}^{2-}$ active ions with increasing temperature. Moreover, it was reported that a chemical mechanism occurs involving interactions between Ca^{2+} and $SO_4{}^{2-}$ and between Mg^{2+} and $SO_4{}^{2-}$ at the chalk surface, which cause displacement of negatively charged carboxylic materials in crude from the positively charged chalk surface. For Ca^{2+} and $SO_4{}^{2-}$ interactions, the adsorption of $SO_4{}^{2-}$ to chalk surface increases with increasing temperature. This leads to a decrease in the positive surface charge, more attachment of Ca^{2+} to chalk surface as the electrostatic repulsion force is decreased and reactions with carboxylic material occur (Zhang et al., 2007). This is according to the equation:

$$\begin{aligned} \mathrm{RCOO^{-}} - \mathrm{Ca} - \mathrm{CaCO_3}(s) + \mathrm{Ca^{2+}} + \mathrm{SO_4^{2-}} \\ = \mathrm{RCOO} - \mathrm{Ca^{+}} + \mathrm{Ca} - \mathrm{CaCO_3}(s) + \mathrm{SO_4^{2-}} \end{aligned} \tag{4.3}$$

In this reaction, $SO_4{}^{2-}$ acts like a catalyst for promoting the increase of Ca^{2+} concentration ions close to the surface. Strand et al. (2008a) observed this mechanism, as with increasing temperature, the retention of $SO_4{}^{2-}$ increases and the concentration of Ca^{2+} decreases. Lager et al. (2007) reported MIE mechanism in carbonate reservoir, where an anion exchange occurs with the adsorbed carboxylate on rock surface. The authors agreed with the Strand et al. (2008a) mechanism of wettability alteration in carbonate rocks using seawater. Nevertheless, they stated that MIE does not require low salinity to decrease oil saturation, as there is no expandable electric double layer. Therefore, they did not expect low salinity water to work in carbonate reservoirs.

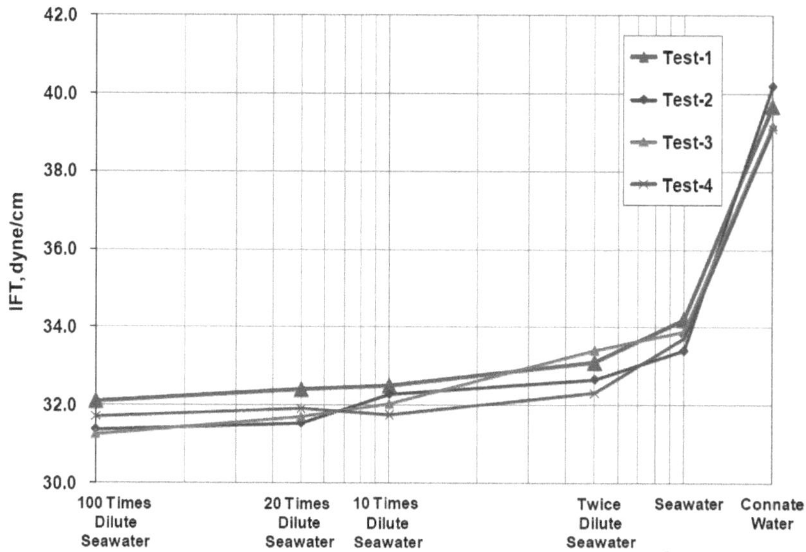

Figure 4.5 IFT measurement of reservoir live-oil with different dilutions of seawater (Yousef et al., 2011).

The key behind increasing oil recovery from chalk reservoir using seawater is related to seawater ions (SO_4^{2-}, Ca^{2+}, and Mg^{2+}) that have the capability of changing rock surface charge, releasing adsorbed carboxylic material to rock surface, altering rock wettability, and eventually improving oil recovery. More pronounced effect of improving oil recovery using high concentration of seawater ions, high temperature formations (>90°C), and sulfate free formation water has been addressed (Yousef et al., 2011).

Moreover, diluting the injected water was reported as a cause for wettability alteration. Yousef et al. (2011) confirmed that the incremental oil recovery due to low salinity water injection is due to wettability alteration, not reduction of IFT. This was possible by measuring both IFT (Fig. 4.5) and contact angle (Fig. 4.6) at reservoir conditions.

Results indicated that IFT has no significant effect on incremental oil recovery due to low salinity water injection and wettability alteration is the main mechanism behind this additional oil increase. They concluded that incremental oil recovery due to low salinity water injection is caused by wettability alteration (fluid–rock interactions) rather than IFT reduction (fluid–fluid interactions). They also proposed that wettability alteration occurs due to change in surface charge and dissolution processes, which was observed through the Nuclear Magnetic Resonance (NMR) technique

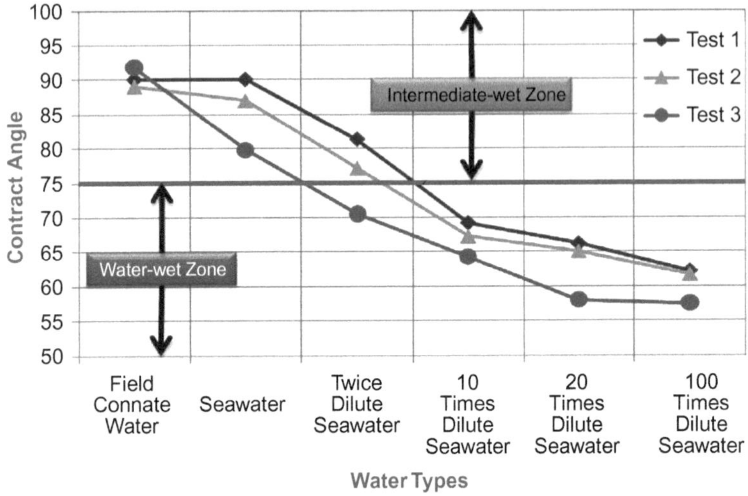

Figure 4.6 Contact angle measurement of reservoir live-oil with different dilutions of seawater (Yousef et al., 2011).

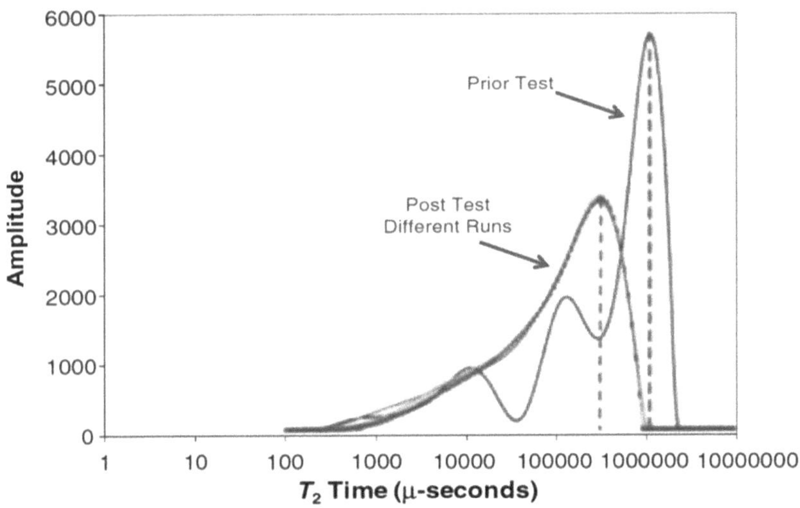

Figure 4.7 NMR T_2 distribution for a rock sample before and after the coreflood experiment (Yousef et al., 2011).

(Fig. 4.7). Left shifting in amplitudes obtained for the post test case as well as overlapping between low and high amplitudes, which means higher rate of relaxation due to change in the carbonate rock surface charge and improvement in pore by dissolution, respectively. This study of smart waterflooding is different from seawater chalk carbonate rock studies in terms of water salinity as diluted water was used and it was not rich with key ions (SO_4^{2-}, Ca^{2+}, and Mg^{2+}) as seawater chalk carbonate rock case.

In the case of rock dissolution, collapsing of some of the pore spaces occurs and, hence, oil is expelled. Pressure, temperature, chemistry of pore water, and oil play a role in wettability alteration and oil recovery by rock dissolution (Hiorth et al., 2010). Tang and Kovscek (2004) reported the decrease in residual oil saturation from field and outcrop diatomite as a result of temperature increase. They also reported that wettability alteration correlated with fines production at high temperature. Schembre and Kovscek (2004) reported that the detachment of oil-wet fines from pore walls exposes the underlying clean water-wet pore surface. Also, Schembre et al. (2006) observed fine migration as a mechanism for wettability alteration from their water imbibition tests. Evje and Hiorth (2009) proposed that the potential ions (SO_4^{2-}, Ca^{2+}, and Mg^{2+}) enter to the rock matrix by molecular diffusion. Then, a nonequilibrium state is created, which results in chemical interactions in the aqueous phases as well as interactions between water and rock in terms of precipitation/dissolution of rock minerals and/or changing the rock surface charge, and, hence, wettability alteration.

In case of injecting water containing Mg^{2+} and SO_4^{2-}, Mg^{2+} has the ability to substitute Ca^{2+} ions from the surface lattice of the chalk. The presence of SO_4^{2-} plays a catalytic role for the substitution process (Zhang et al., 2007). The substitution equation is as follows:

$$\begin{aligned} &RCOO^- - Ca - CaCO_3(s) + Mg^{2+} + SO_4^{2-} \\ &\quad = Mg - CaCO_3(s) + RCOO - Ca^+ + SO_4^{2-} \end{aligned} \tag{4.4}$$

The substitution was observed by Strand et al. (2008b); by flooding seawater through a limestone core at 130°C, they noticed that concentration of Ca^{2+} increased, that concentration of Mg^{2+} decreased, and that concentration of SO_4^{2-} was not much affected as the precipitation of $CaSO_4$(s) was small. Water weakening effect of carbonate rocks was reported in the literature by adding SO_4^{2-} and Mg^{2+} to the imbibing fluid, causing a decrease in the yield point of the chalk rock and dramatic compaction at high temperatures. This means that the chemical interaction between components of seawater and the chalk rock affects the mechanical strength of the chalk. The weakening effect of imbibing water is related to both $CaCO_3$ solubility and substitution of Ca^{2+} by Mg^{2+}. The presence of Mg^{2+} along with SO_4^{2-} enhances the substitution of

Ca^{2+} by Mg^{2+}. The difference of sizes between Ca^{2+} and Mg^{2+} causes stress changes and reduction in the mechanical strength of the chalk rock. This is added to the formed $MgCO_3$ that is more soluble in water compared to $CaCO_3$ (Austad et al., 2008).

Later, Yousef et al. (2012a) conducted more work using Nuclear Magnetic Resonance (NMR), contact angle measurement, and zeta potential studies to demonstrate that wettability alteration is the reason behind LSWI. The results showed that wettability alteration occurs through changing the surface charge from the zeta potential measurements and dissolution of $CaSO_4$ from NMR tests. Moreover, the study addressed the importance of multivalent ions in enhancing wettability alteration. They studied the rock surface chemistry by measuring zeta potential, which describes the magnitude of charge present on a charge particle. Charge particles with high zeta potential are self-stabilizing. The change in surface charge of carbonate rocks is supported by NMR tests showing fast surface relaxation and zeta potential measurements showing shifting towards more negative state with successive injection of seawater dilutions. Further, Yousef et al. (2012b) investigated the applicability of LSWI in both secondary and tertiary modes. The results confirmed the effect of LSWI in the tertiary mode as about 14% of OOIP was obtained by injection of seawater, twice-diluted, 10-times diluted, and 100-times diluted in succession. Moreover, the study showed the potential of LSWI in the secondary mode as 10% additional oil recovery was obtained using 10-times diluted seawater compared to the seawater. Also, the temperature effect on wettability alteration was highlighted by comparing the results of this study with their previous work of Yousef et al. (2011).

Romanuka et al. (2012) reported two ways for modifying the wettability of carbonate rocks using LSWI/EWI: Increasing the concentration of surface interacting ions (SO_4^{2-}, BO_3^{3-} or PO_4^{3-}) in the injected brine and lowering ionic strength of the injected brine. Spontaneous imbibition experiments were conducted in the latter study on both reservoir carbonate and Stevns Klint samples. Results showed that an incremental oil recovery of 4–20% OOIP was obtained from reservoir carbonate samples upon reducing the ionic strength of the injected water. Additional oil recovery was obtained from Stevns Klint samples by increasing sulfate ion concentration as no response was noticed with lowering ionic strength of the injected water. Al-Harrasi et al. (2012) provided a direct evidence of low salinity waterflooding effect on oil recovery from Omani carbonate rocks through spontaneous and corefiooding experiments. Wettability alteration was referred to as the reason for low salinity water injection with the observed negligible reduction in IFT. In addition, they observed the LSWI/EWI effect on oil recovery at higher water salinity level than the reported in the literature. Low salinity water injection by lowering the ionic strength has more pronounced effect on oil recovery from oil-bearing zone cores compared to hardening the injected water. Nevertheless, the case is just the opposite for Stevns Klint outcrop chalk cores because they are more responsive to hardening of the injected low salinity water as was reported by Romanuka et al. (2012) through their spontaneous imbibition experiments.

Al-Shalabi (2014) performed an extensive numerical work in the LSWI area, especially in carbonates, including history matching of then published corefloods (Yousef et al., 2011 and 2012b; Chandrasekhar and Mohanty, 2013), modeling of the process, field-scale predictions, sensitivity analysis, and optimization. In light of the extensive conducted work, Al-Shalabi et al. (2015) interpreted the mechanism controlling the LSWI effect on oil recovery from carbonates based on their findings. They reported that the incremental oil recovery by LSWI in carbonates could best be explained as wettability alteration caused by both changes in surface charge and dissolution (Figs. 4.8 and 4.9).

In most of the literature, the emphasis is on pH causing in situ generation of surfactants; however, Al-Shalabi et al. (2015) emphasized

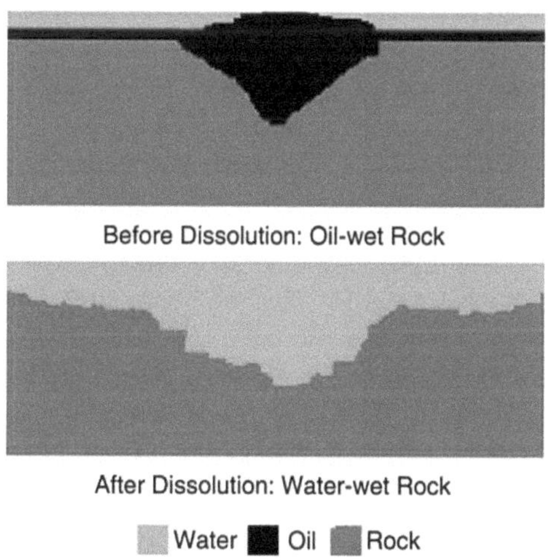

Figure 4.8 Wettability alteration by dissolution (Al-Shalabi et al., 2015).

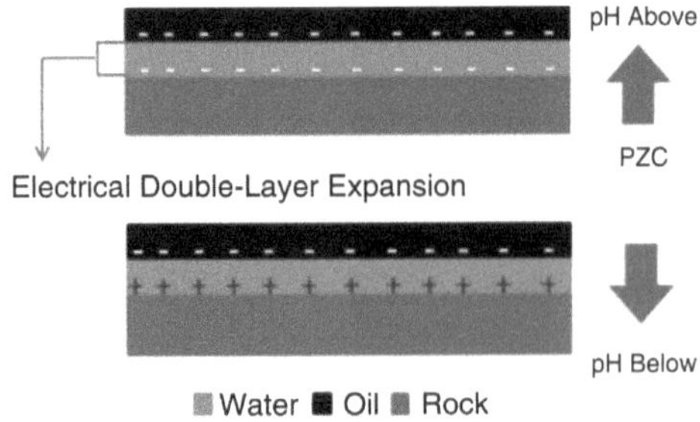

Figure 4.9 pH-induced wettability alteration (Al-Shalabi et al., 2015).

pH-inducted wettability, where the pH of the solution exceeds the point of zero charge (PZC), the rock surface charge changes, the electrical double layer (EDL) expands, the rock wettability alters, and the oil recovery increases. Moreover, they presented a flow chart describing the LSWI mechanism in carbonates (Fig. 4.10).

Alameri et al. (2015) conducted different corefloods at reservoir conditions to investigate the mechanism controlling the incremental oil recovery by LSWI from carbonates. They also conducted contact angle and IFT measurements. Results showed that with the use of LSWI the rock wettability alters from oil-wet to intermediate-wet. Moreover, the use of surfactants further shifts the wettability from intermediate-wet to water-wet depending on the salinity of the injected water. IFT measurements showed an increasing trend with decreasing salinity of injected water, which is not consistent with the observed trend by LSWI. They measured the effluent ionic concentration for one of the experiments, which showed a decrease in Ca^{2+}, Mg^{2+}, Cl^{-}, and SO_4^{2-} ions that might be related to a possible multiions exchange.

Austad et al. (2015) showed that carbonates containing anhydrite have a low salinity effect, which is similar to the one reported previously for seawater (Zhang et al., 2006); however, in this case, the sulfate is provided in situ through anhydrite dissolution. Moreover, they stated that the efficiency of wettability alteration in this case is mainly related to the increase in concertation of sulfate and the decrease in the NaCl concentration at a given temperature.

Mahani et al. (2015) investigated whether calcite dissolution is the primary reason for incremental oil recovery by LSWI. The authors used zeta potential as well as contact angle measurements on planar surfaces coated with crushed limestone and Silurian dolomite rocks. They observed

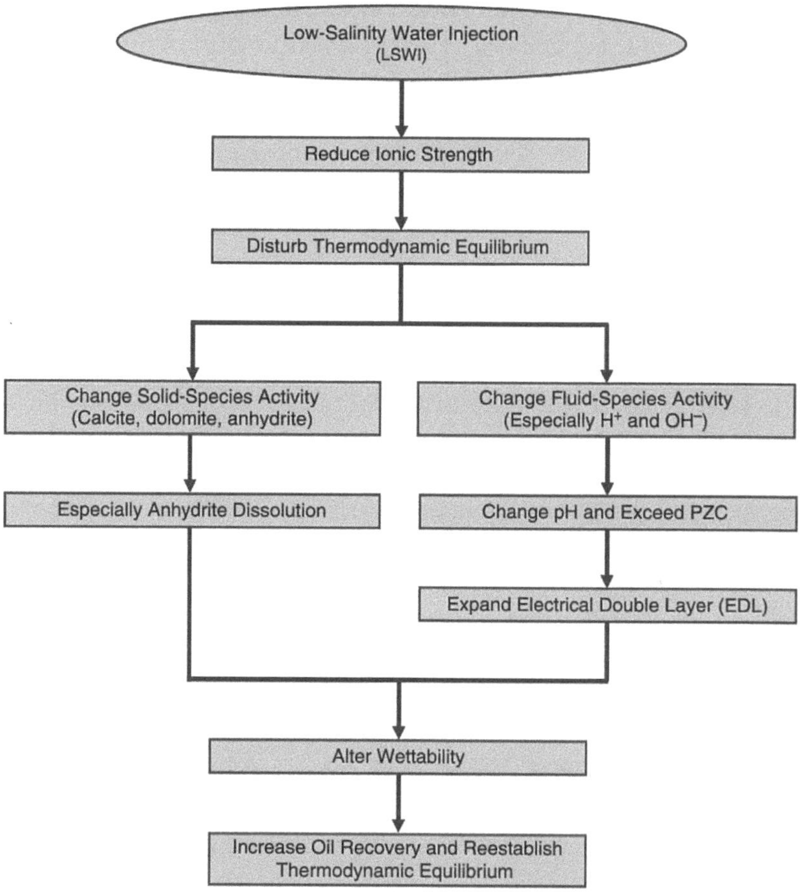

Figure 4.10 Flow chart of LSWI mechanism in carbonates (Al-Shalabi et al., 2015).

decrease in contact angle by switching the injected water from formation water to diluted seawater. Similarly, a decreasing trend of zeta potential was observed with diluting the injected water. Nevertheless, stronger response by LSWI was reported for limestone compared to that of dolomite, which was due to the higher adhesion forces between oil and rock surface in dolomite compared to limestone. Moreover, the authors believe that changing the rock surface charge is the primary mechanism for the low salinity technique, while dissolution is a secondary mechanism, which is consistent with field-scale observations.

In a later study, Mahani et al. (2016) presented a detailed study by measuring the zeta potential for different types of carbonate rocks (calcite, limestone, chalk, and dolomite) over a wide range of salinity, brine composition, and pH. In addition, they performed simulation runs using the surface complexation model implemented in PHREEQC (Parkhurst and Appelo, 2013) in order to relate the calculated surface potentials to the measured zeta potentials. The results showed consistency between the calculated surface potentials using PHREEQC and the measured zeta potentials as both of them increase with increasing the pH of solution and decrease with diluting the injected water or increasing the sulfate concentration. This trend was consistent for all rock types; however, noticeable differences in values were highlighted which stress on the importance of rock minerology.

Jackson et al. (2016) investigated the mechanism controlling incremental oil recovery by low salinity/engineered water in carbonates through zeta potential measurements. They showed that incremental oil recovery by LSWI/EWI is strongly correlated to changes in zeta potential at both mineral-water and oil-water interfaces. Through their experimental work, they highlighted for the first time that the zeta potential at the oil-water interface may be positive at conditions relevant to carbonate reservoirs. The brine composition should be changed in a way which yields a zeta potential at each interface that has the same polarity so that a repulsive electrostatic force acts between the interfaces and stabilizes the water film on the mineral surface. The authors stressed on the need for measuring zeta potential on the intact rock in the presence of reservoir conditions including temperature, pressure, initially high ionic strength formation brine, and oil.

Adegbite et al. (2017) investigated the mechanism behind engineered water injection (EWI) effect on oil recovery through geochemical modeling of carbonate corefloods using the CMG-GEM simulator. In this case the engineered water was a spiked sulfate concertation seawater. Different geochemical reactions were captured in the simulation runs including aqueous, dissolution/precipitations, and ion-exchange reactions. The authors concluded that the wettability alteration is the reason for incremental oil recovery by EWI in carbonates. They believe that wettability alteration is mainly caused by changing of the rock surface charge as well as calcite dissolution. For the change in the rock surface charge that could be either due to pH increase and exceedance of PZC or an anionic exchange reaction between sulfate ion and the carboxylic group in the crude oil. The authors presented a flow chart describing the EWI mechanism in carbonates (Fig. 4.11).

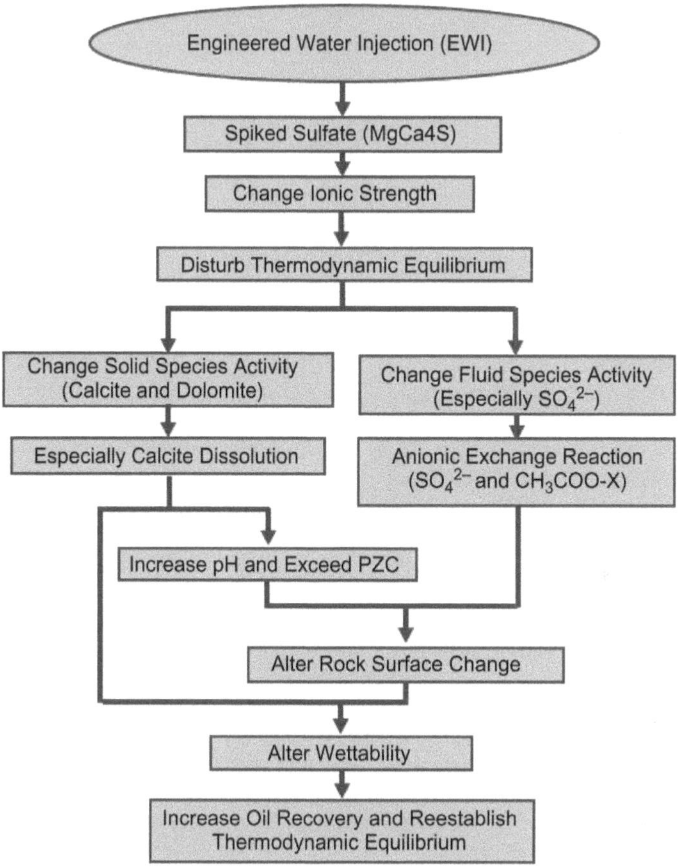

Figure 4.11 Flow chart of EWI mechanism in carbonates (Adegbite et al., 2017).

Usually, the operators try to remove sulfate from the injected water to avoid souring (H_2S production) and scale problems. The presence of sulfate in injection water in high concentration causes scaling problems when combined with high concentration barium or strontium in formation water. Barium sulfate and strontium sulfate scales cause both surface and subsurface problems and affect flood front breakthrough at the producing wells. Scales removal using workover and mechanical techniques is very expensive. Moreover, the use of chemical inhibitors postpones the problem and requires constant monitoring (Healtherly et al., 1994). In addition, reservoir souring by bacterial activity might occur due to sulfate-reducing bacteria (SRB), which produces hydrogen sulfide gas at the producer wells. This problem might be controlled using sulfide scavengers and biocides

(Maxwell and Spar, 2005). Nevertheless, the gain in oil recovery might justify the extra cost of scale inhibitors and sulfide preventers.

As seen from the discussion above, there are different mechanisms controlling the effect of LSWI/EWI on oil recovery for both sandstones and carbonates, and, hence, capturing the primary contributing mechanism is important, especially, in modeling of this technique, which is presented in Chapter 5, Modeling of the LSWI/EWI Technique in Sandstones and Carbonates.

REFERENCES

Adegbite, J.O., Al-Shalabi, E.W., Ghosh, B., 2017. Private communication.

Alameri, W., Teklu, T.W., Graves, R.M., Kazemi, H., AlSumaiti, A.M., 2015. Experimental and numerical modeling of low-salinity waterflood in a low permeability carbonate reservoir. SPE Western Regional Meeting, Garden Grove, California, USA, Paper SPE 174001.

Al-Harrasi, A.S., Al Maamari, R.S., Masalmeh, S., 2012. Laboratory investigation of low salinity waterflooding for carbonate reservoirs. Abu Dhabi International Petroleum Exhibition & Conference, U.A.E, Paper SPE 161468.

Alotaibi, M.B., Nasr-El-Din, H.A., 2009. Chemistry of injection water and its impact on oil recovery in carbonate and clastic formations. SPE International Symposium on Oilfield Chemistry, The Woodlands, Texas, USA, Paper SPE 121565.

Al-Shalabi, E.W., 2014. Modeling the effect of injecting low salinity water on oil recovery from carbonate reservoirs. PhD Dissertation, The University of Texas at Austin, Texas, USA.

Al-Shalabi, E.W., Sepehrnoori, K., Pope, G., 2015. Geochemical interpretation of low salinity water injection in carbonate oil reservoirs. SPE Journal. 20 (6), 1212–1226.

Austad, T., 2008. Smart water for enhance recovery: A comparison of mechanisms in carbonates and sandstones. Force RP Work Shop: Low Salinity Water Flooding, the Importance of Salt Content in Injection Water, Norway.

Austad, T., Strand, K., Madland, M.V., Puntervold, T., Korsnes, R.I., 2008. Sea water in chalk: An EOR and compaction fluid. SPE Reserv. Eval. Eng. 11 (4), 648–654.

Austad, T., RezaeiDoust, A., Puntervold, T., 2010. Chemical mechanism of low salinity water flooding in sandstone reservoirs. SPE Improved Oil Recovery Symposium, Tulsa, Oklahoma, USA, Paper SPE 129767.

Austad, T., Shariatpanahi, S.F., Strand, S., Aksulu, H., Puntervold, T., 2015. Low salinity EOR-effects in limestone reservoir cores containing anhydrite: A discussion of the chemical mechanism. Energ. Fuel. 29 (11), 6903–6911.

Bernard, G.G., 1967. Effect of floodwater salinity on recovery of oil from cores containing clays. SPE California Regional Meeting, Los Angeles, California, USA, Paper SPE 1725.

Chandrasekhar, S., Mohanty, K.K., 2013. Wettability alteration with brine composition in high temperature carbonate reservoirs. SPE Annual Technical Conference and Exhibition, New Orleans, Louisiana, USA, Paper SPE 166280.

Computer Modeling Group (CMG), 2016. User technical manual.

Doust, A.R., Puntervold, T.P., Strand, S., Austad, T.A., 2009. Smart water as wettability modifier in carbonate and sandstone. 15th European Symposium on Improved Oil Recovery, Paris, France.

Doust, A.R., Puntervold, T.P., Austad, T., 2010. A discussion of the low salinity EOR potential for a North Sea sandstone field. SPE Annual Technical Conference and Exhibition, Florence, Italy, Paper SPE 134459.

Evje, S., Hiorth, A., 2009. A mathematical model for dynamic wettability alteration controlled by water-rock chemistry. Netw. Heterog. Media. 5 (2), 217–256.

Hassenkam, T., Matthiesen, J., Pedersen, C.S., Dalby, K.N., Stipp, S.L.S., Collins, I.R., 2012. Observation of the low salinity effect by atomic force adhesion mapping on reservoir sandstones. SPE Improved Oil Recovery Symposium, Tulsa, OK, USA, Paper SPE 154037.

Hassenkam, T., Andersson, H., Hilner, E., Matthiesen, J., Dobberschutz, S., Dalby, K.N., et al., 2014. A fast alternative to core plug tests for optimizing injection water salinity for EOR. SPE Improved Oil Recovery Symposium, Tulsa, OK, USA, Paper SPE 169136.

Healtherly, M.W., Howell, M.E., McElhiney, J.E., 1994. Sulfate removal technology for seawater waterflood injection. Offshore Technology Conference, Houston, Texas, USA, Paper SPE 7593.

Hiorth, A., Cathles, L.M., Madland, M.V., 2010. Impact of pore water chemistry on carbonate surface charge and oil wettability. Transport Porous Med. 85 (1), 1–21.

Hognesen, E.J., Strand, S., Austad, T., 2005. Waterflooding of preferential oil-wet carbonates: Oil recovery related to reservoir temperature and brine composition. SPE EUROPEC/EAGE Annual Conference, Madrid, Spain, Paper SPE 94166.

Jackson, M.D., Al-Mahrouqi, D., Vinogradov, J., 2016. Zeta potential in oil-water-carbonate systems and its impact on oil recovery during controlled salinity waterflooding. Sci. Rep. 6, 1–13, Article 37363.

Lager, A., Webb, K.J., Black, C.J.J., Singleton, M., Sorbie, K.S., 2006. Low salinity oil recovery-an experimental investigation. Proceedings of International Symposium of the Society of Core Analysts, Norway.

Lager, A., Webb, K.J., Black, C.J.J., 2007. Impact of brine chemistry on oil recovery. 14th European Symposium on IOR, Cairo, Egypt.

Lager, A.K., Webb, K.J., Collins, I.R., Richmond, D.M., 2008. LoSal™ enhanced oil recovery: Evidence of enhanced oil recovery at the reservoir scale. SPE Symposium on Improved Oil Recovery, Tulsa, Oklahoma, USA, Paper SPE 113976.

Lee, S.Y., Webb, K.J., Collins, I., Lager, A., Clarke, S., O'Sullivan, M., et al., 2010. Low salinity oil recovery: Increasing understanding of the underlying mechanisms. SPE Improved Oil Recovery Symposium, Tulsa, Oklahoma, USA, Paper SPE 129722.

Ligthelm, D.J., Gronsveld, J., Hofman, J., Brussee, N., Marcelis, F., Linde, H.V.D., 2009. Novel waterflooding strategy by manipulation of injection brine composition. EUROPEC/EAGE Conference and Exhibition, Amsterdam, The Netherlands, Paper SPE 119835.

Mahani, H., Berg, S., llic, D., Bartels, W.B., Niasar, V.J., 2014. Kinetics of low-salinity-flooding effect. SPE Journal. 20 (1), 8–20.

Mahani, H., Keya, A.L., Berg, S., Bartels, W., Nasralla, R., Rossen William, 2015. Driving mechanism of low salinity flooding in carbonate rocks. SPE EUROPEC, Madrid, Spain, Paper SPE 174300.

Mahani, H., Keya, A.L., Berg, S., Nasralla, R., 2016. Electrokinetics of carbonate/brine interface in low-salinity waterflooding: Effect of brine salinity, composition, rock type, and pH on zeta-potential and a surface-complexation model. SPE Journal. In press.

Martin, J.C., 1959. The effects of clay on the displacement of heavy oil by water. Venezuelan Annual Meeting, Caracas, Venezuela, Paper SPE 1411.

Maxwell, S., Spar, I., 2005. Souring of reservoirs by bacterial activity during seawater waterflooding. SPE International Symposium on Oilfield Chemistry, The Woodlands, Texas, USA, Paper SPE 93231.

McGuire, P.L., Chatham, J.R., Paskvan, F.K., Sommer, D.M., Carini, F.H., 2005. Low salinity oil recovery: An exciting new EOR opportunity for Alaska's North Slope. SPE Western Regional Meeting, Irvine, California, USA, Paper SPE 93903.

Nasralla, R.A., Nasr-El-Din, H.A., 2011. Impact of electrical surface charges and cation exchange on oil recovery by low salinity water. SPE Asia Pacific Oil and Gas Conference and Exhibition, Jakarta, Indonesia, Paper SPE 147937.

Nasralla, R.A., Nasr-El-Din, H.A., 2012. Double-layer expansion: Is it a primary mechanism of improved oil recovery by low-salinity waterflooding? SPE Improved Oil Recovery Symposium, Tulsa, Oklahoma, USA, Paper SPE 154334.

Nasralla, R.A., Alotaibi, M.B., Nasr-El-Din, H.A., 2011a. Efficiency of oil recovery by low salinity water flooding in sandstone reservoirs. SPE Western North American Region Meeting, Alaska, USA, Paper SPE 144602.

Nasralla, R.A., Bataweel, M.A., Nasr-El-Din, H.A., 2011b. Investigation of wettability alteration by low salinity water in sandstone rock. Offshore Europe Meeting, Aberdeen, UK, Paper SPE 146322.

Parkhurst, D.L., Appelo, C.A.J., 2013. Description of input and examples for PHREEQC Version 3—a computer program for speciation, batch-reaction, one-dimensional transport, and inverse geochemical calculations. US Geological Survey Techniques and Methods (Chapter 43 of Section A Groundwater, Book 6 Modeling Techniques, p. 497).

Puntervold, T., Strand, S., Austad, T., 2007. Waterflooding of carbonate reservoirs: Effects of a model base and natural crude oil bases on chalk wettability. Energ. Fuel. 21 (3), 1606–1616.

Rivet, S., Lake, L.W., Pope, G.A., 2010. A coreflood investigation of low-salinity enhanced oil recovery. SPE Annual Technical Conference and Exhibition, Florence, Italy, Paper SPE 134297.

Romanuka, J., Hofman, J.P., Ligthelm, D.J., Suijkerbuijk, B.M.J.M., Marcelis, A.H.M., Oedai, S., et al., 2012. Low salinity EOR in carbonates. SPE Improved Oil Recovery Symposium, Tulsa, Oklahoma, USA, Paper SPE 153869.

Rotondi, M., Callegaro, C., Masserano, F., Bartosek, M., 2014. Low salinity water injection: Eni's experience. Abu Dhabi International Petroleum Exhibition and Conference, Abu Dhabi, UAE, Paper SPE 171794.

Schembre, J.M., Kovscek, A.R., 2004. Thermally induced fines mobilization: Its relationship to wettability and formation damage. SPE International Thermal Operations and Heavy-Oil Symposium, and Western Regional Meeting, California, USA, Paper SPE 86937.

Schembre, J., Tang, G.Q., Kovscek, A., 2006. Wettability alteration and oil recovery by water imbibition at elevated temperatures. J. Petrol. Sci. Eng. 52 (1–4), 131–148.

Shehata, A.H., Nasr-El-Din, H.A., 2015. Zeta potential measurements: Impact of salinity on sandstone minerals. SPE International Symposium on Oilfield Chemistry, The Woodlands, Texas, USA, Paper SPE 173763.

Smith, K.W., 1942. Brines as flooding liquids. Seventh Annual Technical Meeting. Mineral Industries Experiment Station, Pennsylvania State College, Pennsylvania, USA.

Sohrabi, M., Mahzari, P., Farzaneh, S.A., Mills, J.R., Tsolis, P., Ireland, S., 2015. Novel insights into mechanisms of oil recovery by low salinity water injection. SPE Middle East Oil & Gas Show and Conference, Manama, Bahrain, Paper SPE 172778.

Standnes, D.C., Austad, T., 2000. Wettability alteration in chalk: 2. Mechanism for wettability alteration from oil-wet to water-wet using surfactants. J. Petrol. Sci. Eng. 28 (3), 123–143.

Strand, S., Standnes, D.C., Austad, T., 2003. Spontaneous imbibition of aqueous surfactant solution into neutral to oil wet carbonate cores: Effects of brine salinity and composition. Energ. Fuel. 17 (5), 1133–1144.

Strand, S., Puntervold, T., Austad, T., 2008a. Effect of temperature on enhanced oil recovery from mixed wet chalk cores by spontaneous imbibition and forced displacement using seawater. Energ. Fuel. 22 (5), 3222–3225.

Strand, S., Austad, T., Puntervold, T., Hognesen, E.J., Olsen, M., Barstad, S.M.F., 2008b. Smart water for oil recovery from fractured limestone: A preliminary study. Energ. Fuel. 22 (5), 3126–3133.

Suijkerbuijk, B.M.J.M., Hofman, J.P., Ligthelm, D.J., Romanuka, J., Brussee, N., van der Linde, H.A., et al., 2012. Fundamental investigations into wettability and low salinity flooding by parameter isolation. SPE Improved Oil Recovery Symposium, Tulsa, Oklahoma, USA, Paper SPE 154204.

Tang, G., Kovscek, A.R., 2004. An experimental investigation of the effect of temperature on recovery of heavy-oil from diatomite. SPE Journal. 9 (2), 163–179.

Tang, G.Q., Morrow, N.R., 1997. Salinity temperature, oil composition and oil recovery by waterflooding. SPE Reserv. Eng. 12 (4), 269–276.

Tang, G.Q., Morrow, N.R., 1999. Influence of brine composition and fines migration on crude oil/brine/rock interactions and oil recovery. J. Petrol. Sci. Eng. 24 (2–4), 99–111.

Yousef, A.A., Al-Saleh, S., Al-Kaabi, A., Al-Jawfi, M., 2011. Laboratory investigation of the impact of injection-water salinity and ionic content on oil recovery from carbonate reservoirs. SPE Reserv. Eval. Eng. 14 (5), 578–593.

Yousef, A.A., Al Saleh, S., Al Jawfi, M., 2012a. The impact of the injection water chemistry on oil recovery from carbonate reservoirs. SPE EOR Conference at Oil and Gas West Asia, Muscat, Oman, Paper SPE 154077.

Yousef, A.A., Al Saleh, S., Al Jawfi, M., 2012b. Improved/enhanced oil recovery from carbonate reservoirs by tuning injection water salinity and ionic content. SPE Improved Oil Recovery Symposium, Tulsa, Oklahoma, USA, Paper SPE 154076.

Zhang, P., Tweheyo, M.T., Austad, T., 2006. Wettability alteration and improved oil recovery in chalk: The effect of calcium in the presence of sulfate. Energ. Fuel. 20 (5), 2056–2062.

Zhang, P., Tweheyo, M.T., Austad, T., 2007. Wettability alteration and improved oil recovery by spontaneous imbibition of seawater into chalk: Impact of the potential determining ions Ca^{2+}, Mg^{2+}, and SO_4^{2-}. Colloid. Surface. A. 301 (1–3), 199–208.

Zhang, Y., Morrow, N.R., 2006. Comparison of secondary and tertiary recovery with change in injection brine composition for crude oil/sandstone combinations. SPE Symposium on Improved Oil Recovery, Tulsa, Oklahoma, USA, Paper SPE 99757.

CHAPTER FIVE

Modeling of the LSWI/EWI Technique in Sandstones and Carbonates

Contents

This chapter presents the different techniques used in the numerical modeling of the low salinity/engineered water injection (LSWI/EWI) process.

5.1 GENERAL MODELING APPROACH

Few modeling works for LSWI/EWI exist in the literature because the researchers focused their works on checking the applicability of this technique and understanding the mechanism underlying the related incremental oil recovery. In this section, we highlight the general modeling works of LSWI/EWI for both sandstones and carbonates.

The simplest way of modeling LSWI/EWI is the linear interpolation between two sets of relative permeability curves, capillary pressure curves, and residual oil saturation values representing an oil-wet state (initial condition) and water-wet state (final condition) using a scaling (interpolating) factor as follows:

$$\begin{aligned} k_{rl}^{altered} &= \theta_1 \times k_{rl}^{initial} + (1-\theta_1) \times k_{rl}^{final}, \\ P_c^{altered} &= \theta_2 \times P_c^{initial} + (1-\theta_2) \times P_c^{final}, \\ S_{or}^{altered} &= \theta_3 \times S_{or}^{initial} + (1-\theta_3) \times S_{or}^{final}, \end{aligned} \tag{5.1}$$

where θ_1, θ_2, and θ_3 are interpolating factors, k_{rl} is the relative permeability of the phase l, and P_c is the capillary pressure. Usually, the oil-wet state

Low Salinity and Engineered Water Injection for Sandstone and Carbonate Reservoirs.
DOI: http://dx.doi.org/10.1016/B978-0-12-813604-1.00005-5

represents the high-salinity injection cycle, while the low salinity injection cycle is represented by the water-wet state.

Jerauld et al. (2008) used the latter approach for predicting low salinity water injection oil recovery at field-scale based on corefloods and single-well tests. The proposed model considers salt as an additional single lumped component in the aqueous phase, and the density and viscosity of the aqueous phase, relative permeability, and capillary pressure curves all depend on salinity. Moreover, both clay content effect and dispersion were considered; dispersion was assigned a value of 5% of the system's length. Equations included in the model for representing capillary curves and relative permeability curves with upper and lower salinity limits are

$$
\begin{aligned}
k_{rw} &= \theta \times k_{rw}^{HS}(S^*) + (1-\theta) \times k_{rw}^{LS}(S^*), \\
k_{ro} &= \theta \times k_{ro}^{HS}(S^*) + (1-\theta) \times k_{ro}^{LS}(S^*), \\
P_{cow} &= \theta \times P_{cow}^{HS}(S^*) + (1-\theta) \times P_{cow}^{LS}(S^*), \\
\theta &= \left(S_{orw} - S_{orw}^{LS}\right) / \left(S_{orw}^{HS} - S_{orw}^{LS}\right), \\
S^* &= (S_o - S_{orw})/(1 - S_{wr} - S_{orw}).
\end{aligned}
\tag{5.2}
$$

For investigating the stability of the low salinity waterfront, Tripathi and Mohanty (2008) modeled instability of low salinity water injection floods in the tertiary mode without including wettability alteration effect on capillary pressure. The analytical 1D solution shows two shocks with probability of instability in the rear shock (low salinity shock) due to the adverse mobility ratio. This was confirmed using the viscous fingering theory and 2D numerical simulation model. They concluded that low salinity water injection is expected to be mildly unstable at the rear front.

For conventional and fractured medium, Wu and Bai (2009) proposed mathematical model for wettability alteration modeling due to low salinity waterflooding. In this model, salt is treated as an additional component in the aqueous phase, which is transported by advection and diffusion, including the adsorption effect on the surface of the rock. Improving oil recovery was incorporated by treating relative permeability, capillary pressure, and residual oil saturations as salinity dependent.

The importance of dynamic wettability alteration models is highlighted in the literature. Spontaneous imbibition experiments of Stevns Klint chalk core plugs using low salinity water injection were simulated using a 1D model proposed by Yu et al. (2009). The model accounts for molecular diffusion, salt adsorption, gravitational, and capillary forces for simulating the dynamic wettability alteration process. The dynamic changes in both capillary pressure and relative permeability curves from oil-wet to water-wet were made dependent on wettability alteration agent or salt concentration. In addition, the salt concentration in the model was considered as $c_a + w_a = 1$ (mole fraction), where c_a is the mass fraction of salt component in aqueous phase, and w_a is the mass fraction of water component in aqueous phase. Moreover, the adsorption isotherm of salt onto rock (c_r) was introduced and made a function of salt concentration, $c_r = f\ (c_a)$. The Langmuir-type isotherm relation for adsorption is used to represent the dependency of salt adsorption (c_r) on salt concentration (c_a) with constants $a_1,\ a_2 > 0$. The effect of wettability alteration on capillary pressure and relative permeability was considered by interpolation between the oil-wet and water-wet cases as follows:

$$\begin{aligned} k_{rl} &= F \times k_{rl}^{ow} + (1 - F) \times k_{rl}^{ww}, \\ P_c &= F \times P_c^{ow} + (1 - F) \times P_c^{ww}, \\ F(c_r) &= \frac{(a^* - c_r)}{a^*}, \\ a^* &= \frac{a_1}{a_2}, \\ c_r &= \frac{a_1 c_a}{(1 + a_2 c_a)}. \end{aligned} \tag{5.3}$$

Good history match was obtained between experimental data and simulation data with emphasis on the importance of using dynamic wettability alteration compared with fixed wettability alteration. For dynamic case, the capillary pressure depends on salt concentration adsorbed onto rock surface and gradually changes from oil-wet to water-wet. Nevertheless, in fixed wettability alteration case, the final wettability altered capillary pressure curve is used directly, independent of salt concentration, which leads to overestimated oil recovery. Alameri et al. (2015) conducted experimental corefloods and modeled them numerically using a 1D two-phase Buckley–Leverett model. The experimental work was history matched successfully using their numerical model with adjusting relative permeability functions to account for wettability alteration.

For capturing fine migration effect on oil recovery, Lemon et al. (2011) provided a simple analytical fines migration model for justifying incremental oil recovery due to low salinity water injection. The model was proposed by combining a modified particle-detachment model along with Dietz model for waterflooding in a layer-cake reservoir. This model was validated for single flow phases by comparing laboratory coreflooding experiments. They reported that fines migration effect is more pronounced when viscosity ratio and heterogeneity of the reservoir increase, but the model still requires validation for two-phase oil/water system. Aladasani et al. (2012a) reported that S_{or} prediction models have poor confidence level unless they consider clay content values and a wettability index. In addition, the reservoir simulations in the same study showed that the incremental oil recovery depends on initial and final wetting states. At strong water-wet conditions, the increase of oil relative permeability is the underlying recovery mechanism, whereas at weak water-wet conditions, the incremental oil recovery is driven by low capillary pressure. Moreover, at both weak and strong oil-wet conditions, oil relative permeability controls additional oil recovery.

Aladasani et al. (2012b) reported that intermediate wetting condition is the ideal wettability alteration condition as capillary pressure is low and the increase in oil relative permeability is the main controller of recovery mechanism. Moreover, the validity of the simulator and recovery mechanism presented were confirmed by matching Yousef et al. (2011) experimental work. This was achieved by considering a linear relationship between salt concentration and each of residual oil saturation, contact angle, and IFT. Nevertheless, some of the disadvantages of this history match are the following: The original laboratory data were not shown in matching the data, the overall recovery values were not exactly in match, the initial jumps in oil recovery were not properly matched, pressure drop data were not matched, and the dependence of residual oil saturation on trapping number was not considered due to exceeding the critical trapping number value.

A systematic way for history matching LSWI corefloods is also presented in the literature. Al-Shalabi et al. (2014a and 2014b) presented a systematic way of history matching published LSWI corefloods, including Yousef et al. (2011 and 2012) and Chandrasekhar and Mohanty (2013) (Table 5.1). The study highlighted the sensitivity of residual oil saturation, capillary pressure curve, and relative permeability parameters (endpoints and Corey's exponents) on LSWI through history matching both oil recovery and pressure drop data for the corefloods investigated. It is worth mentioning that the previously discussed LSWI/EWI models have treated the water and oil phases similarly with respect to capillary pressure (P_c) and relative permeability (k_r), in the sense that both phases are shifted using a similar scaling factor. Al-Shalabi et al. (2015a and 2015b) showed

Table 5.1 Summary of the six proposed LSWI history matching methods (Al-Shalabi, 2014)

Method	Description
First method (S_{or} contribution)	Change S_{or}, keep k_r parameters same as seawater cycle
Second method (S_{or} and k_r contributions)	
First approach	Change Corey's exponents, keep endpoint relative permeability same as seawater cycle
Second approach	Change endpoint relative permeabilities, keep Corey's exponents same as seawater cycle
Third approach	Change Corey's exponent and endpoint relative permeabilities
Third method	
Single ωkn	History matching was not possible
$\omega kn1$ (water) and $\omega kn2$ (oil)	History matching was performed successfully
Fourth method	Simple S_{or} linear interpolation model (ωS)
Fifth method	Water relative permeability parameters are similar to the seawater cycle
Sixth method	Average constant water relative permeability parameters are utilized from wettability alteration cycles

that oil relative permeability is more sensitive to LSWI compared with water relative permeability, and, hence, water and oil phases should be treated separately.

Based on the latter finding, they proposed several models for the effect of LSWI on oil recovery from carbonates that are discussed briefly. It is worth mentioning that the same approach can be applied on sandstones. It should be noted that the simulation and modeling work was performed using the UTCHEM simulator, a 3D nonisothermal chemical compositional flow simulator developed at The University of Texas at Austin (UTCHEM Technical Documentation, 2000). Moreover, the capillary pressure contributions were neglected in the proposed models due to the following: Different capillary pressure curves were assumed in the simulation runs with minimal effect, the researchers used relatively long cores, and the applied high pressure gradient in the laboratory for the corefloods investigated.

An empirical LSWI model was proposed by Al-Shalabi et al. (2014c) and implemented in the UTCHEM simulator. The model captures the change in residual oil saturation, oil endpoint relative permeability, and oil exponent while maintaining constant water relative permeability curve. The assumption of using constant water relative permeability parameters is based on the negligible change in water relative permeability parameters for the corefloods investigated (Yousef et al., 2011 and 2012; Chandrasekhar and Mohanty, 2013). In case experimental data are available, an average constant water relative permeability values can be considered for the seawater dilution cycles; otherwise, the water relative permeability values can be assumed constant and equal to the seawater injection cycle. In this model, residual oil saturation, oil relative permeability endpoint, and oil relative permeability exponent are functions of contact angle measurements, upon which a third degree polynomial function is used to express contact angle as a function of total injected water salinity. Initial and injected water salinities are specified as input parameters. The injected brine is mixed with the initial reservoir brine, and during the mixing process, the salinity wave moves and changes the contact angle in each gridblock using the proposed polynomial function. Once the contact angle changes, then residual oil saturation, oil relative permeability exponent, and oil relative permeability endpoint are calculated using the proposed correlations. The equations implemented in the UTCHEM simulator are as follows:

$$S_{or(Altered)} = \omega S \times S_{or}^{LS} + (1 - \omega S) \times S_{or}^{HS}, \tag{5.4}$$

$$\omega S = \frac{(\theta - \theta^{HS})}{(\theta^{LS} - \theta^{HS})}, \tag{5.5}$$

$$k_{ro}^{*} = \frac{k_{ro}^{*LS} - k_{ro}^{*HS}}{1 + \left(\frac{\theta}{a}\right)^{e}} + k_{ro}^{*HS}, \tag{5.6}$$

$$n_o = \frac{n_{o\max} - n_o^{LS}}{1 + \left(\frac{\theta}{a}\right)^{-e}} + n_o^{LS}. \tag{5.7}$$

Examples of the history matched cumulative oil recovery data for Yousef et al. (2011), shown previously in Fig. 2.2 (Chapter 2), and the relative permeability curves used are depicted in Figs. 5.1 and 5.2, respectively.

In a later work, Al-Shalabi et al. (2014d) presented a model that captures the main effect of LSWI on improving the microscopic displacement efficiency through modeling residual oil saturation as a function of

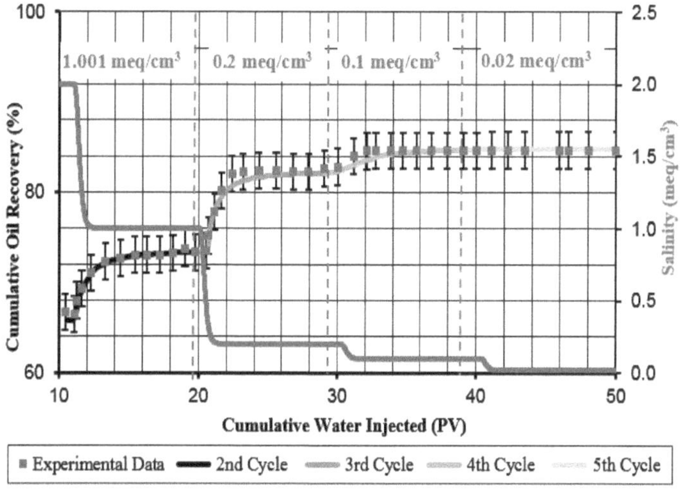

Figure 5.1 Cumulative oil recovery match with the empirical LSWI model (Al-Shalabi et al., 2014c).

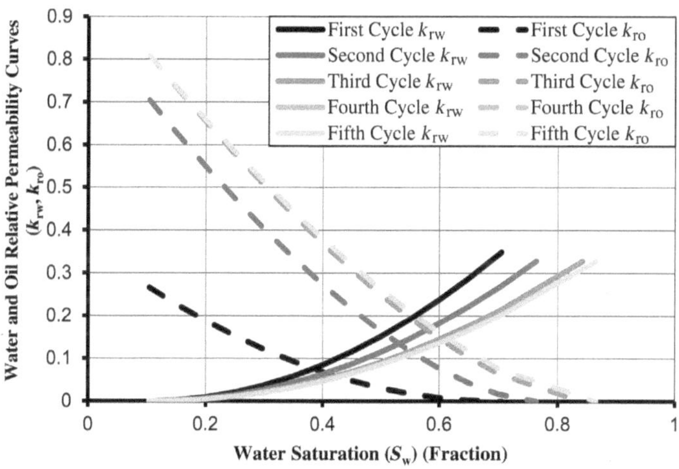

Figure 5.2 Relative permeability curves with the empirical LSWI model (Al-Shalabi et al., 2014c).

trapping number (N_T). This model is named as the Fundamental LSWI Model. The model captures the fundamental microscopic displacement efficiency through adjusting capillary desaturation curve (CDC) using different trapping numbers and trapping parameters (T_l). The main features of the model are the following: Contact angle is a function of the EDL thickness, trapping parameter is a function of contact angle, residual oil saturation is a function of trapping number, and oil relative permeability parameters are functions of the term $TN_T^{\tau *}$ while maintaining constant water relative permeability parameters. The proposed model considers total/stoichiometric ionic strength (I) of the solution, which can be calculated as follows:

$$I = \frac{1}{2}\sum_i z_i^2 m_i, \tag{5.8}$$

where z_i is the charge of the fluid species i and m_i is the molality of fluid species i in moles per kilogram of water. The calculated ionic strength can be used to calculate the thickness of the electric double layer (EDL) or the Debye length (κ^{-1}) as follows (Stumm and Morgan, 1996):

$$\kappa^{-1} = \sqrt{\frac{\varepsilon_r \varepsilon_0 k_B T}{2N_A e^2 I}}, \tag{5.9}$$

where ε_r is the relative permittivity (55.42 for water at reservoir temperature), ε_0 is the permittivity of free space (8.854×10^{-12} $\mathrm{F \cdot m^{-1}}$), k_B is the Boltzmann constant ($1.381 \times 10^{-23}\ \mathrm{J \cdot K^{-1}}$), T is the reservoir temperature, N_A is the Avogadro constant ($6.022 \times 10^{+23}$ 1/mole), e is the elementary charge (1.602×10^{-19} C), and I is the total/stoichiometric ionic strength (mole/m^3). This approach states that as the salinity of the injected water decreases, the ionic strength of the solution decreases and the EDL thickness increases causing a decrease in the contact angle toward more water-wet. Some of the related equations are listed below (Pope et al., 2000; Jin, 1995):

$$S_{lr} = S_{lr}^{high} + \frac{S_{lr}^{low} - S_{lr}^{high}}{1 + T_l (N_{Tl}{}^{\tau_l^*})}, \tag{5.10}$$

$$N_{T_l} = \frac{\left| \vec{k} . \left[\nabla \Phi_{l'} + g(\rho_{l'} - \rho_l) \nabla D \right] \right|}{\sigma_{ll'}}, \tag{5.11}$$

$$\theta = A_t + \frac{B_t}{\kappa^{-1}}, \tag{5.12}$$

where θ is the contact angle in degrees, A_t and B_t are fitting parameters, κ^{-1} is the EDL thickness in nm, τ^* incorporates both heterogeneity and initial oil saturation effects on residual oil saturation, l is the displaced phase, l' is the displacing phase, $\Delta\Phi_{l'}$ is the gradient of the flow potential, k is the permeability, g is the gravitational force constant, and $\sigma_{ll'}$ is the interfacial tension between the displacing and displaced phases. As an example, Fig. 5.3 shows

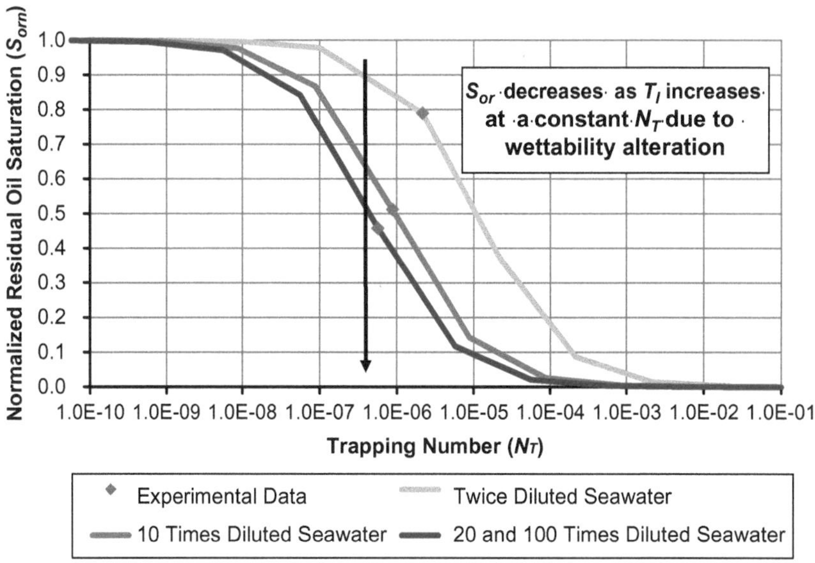

Figure 5.3 CDC model used in the Fundamental LSWI Model (Al-Shalabi et al., 2014d).

the capillary desaturation curve model of Yousef et al. (2011), where residual oil saturation decreases as trapping parameter increases at a constant trapping number, which reflects rendering the rock wettability more water-wet.

For each of the proposed models, there is a range of applicability and screening criteria to follow. Al-Shalabi (2014) addressed the screening criteria for the proposed models. The author believes that no universal LSWI model can be obtained for the following reasons:

- Rock type (sandstone, carbonate) matters, as the rock type affects the wettability alteration process leading to substantial oil recovery.
- Oil recovery mode (secondary, tertiary) is also important.
- Brine injection mode plays a role in the interactions occurring in the oil–water–rock system, which includes LSWI by reducing the overall salinity level (ionic strength) and EWI by both softening and hardening of the injected water.
- Initial wettability state of the rock matters (oil-wet, weakly-oil-wet, neutral-wet, weakly-water-wet, and water-wet) as the LSWI effect is more pronounced in oil-wet rocks.

From the previous points, the proposed models address the following conditions:

- Carbonate rock type

- Secondary and tertiary oil recovery modes
- Reducing overall salinity level as the brine injection mode (LSWI)
- Weakly-oil-wet to mixed-wet rocks for the initial wettability state

5.2 FIELD-SCALE MODELING AND OPTIMIZATION OF LSWI/EWI

For field-scale applications of LSWI, Al-Shalabi et al. (2015c) investigated the effect of LSWI on both oil displacement and volumetric sweep efficiencies using a quarter 5-spot field model, which was created by upscaling the coreflood of Yousef et al. (2011). Both fractional flow and tracer methods were consistent in estimating the volumetric sweep efficiency due to LSWI. An example of the quarter 5-spot field model showing the improvement in volumetric sweep efficiency using LSWI is depicted in Fig. 5.4.

Also, an example of the fractional flow curves showing the improvement in displacement sweep efficiency using LSWI is depicted in Fig. 5.5.

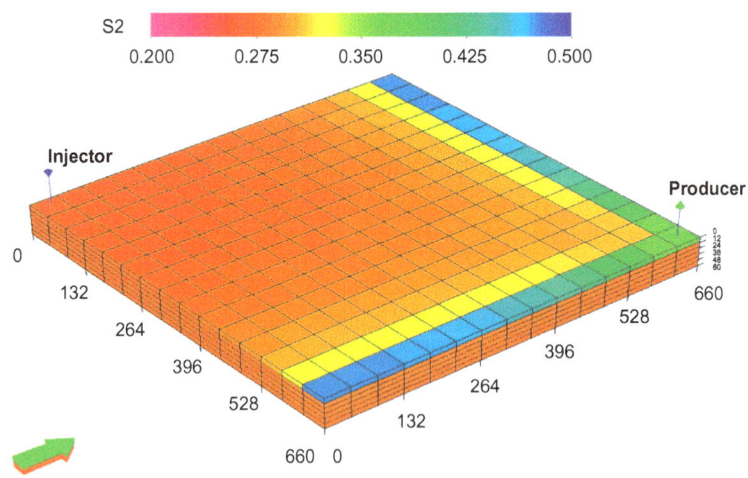

Figure 5.4 3D map of remaining oil saturation at 6 injected PV (Quarter 5-spot Model-LSWI Cycle) (Al-Shalabi et al., 2015c).

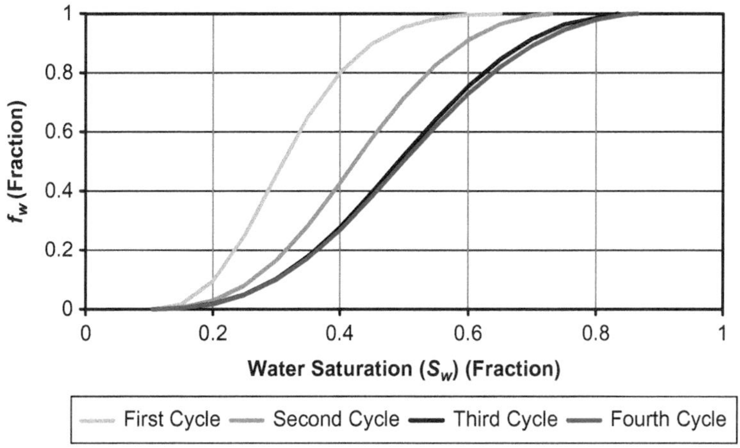

Figure 5.5 Fractional flow curves (empirical LSWI model-quarter 5-spot field model) (Al-Shalabi et al., 2015c).

Craig (1993) showed that areal sweep efficiency is best correlated with the average saturation mobility ratio ($\overline{M}$), which is given by

$$\overline{M} = \frac{(\lambda_{r1} + \lambda_{r2})\Big|_{S_1 = \overline{S}_1}}{(\lambda_{r1} + \lambda_{r2})\Big|_{S_1 = S_{1_I}}}, \tag{5.13}$$

where λ_{r1} is the relative mobility of water, λ_{r2} is the relative mobility of oil, $\overline{S}_1$ is the average water saturation, and S_{1I} is the initial water saturation. The average mobility ratio is the ratio of the total relative mobility at the average water saturation behind the shock front to the total relative mobility at the initial water saturation. Up to date, the sweep efficiency is still correlated with the average saturation mobility ratio ($\overline{M}$) although physically the average saturation is neither existing nor accurate. Al-Shalabi et al. (2015c) proposed a new definition of mobility ratio at the tertiary mode that best correlates with volumetric sweep efficiency and physically makes sense. The proposed definition of mobility ratio is as the following:

$$M = \frac{(\lambda_{r1})\Big|_{S_1 = S_{wf}}}{(\lambda_{r1} + \lambda_{r2})\Big|_{S_1 = S_{OB}}}. \tag{5.14}$$

In this case, the mobility ratio is defined as the ratio of the relative water mobility at the shock front to the total relative mobility at the oil bank

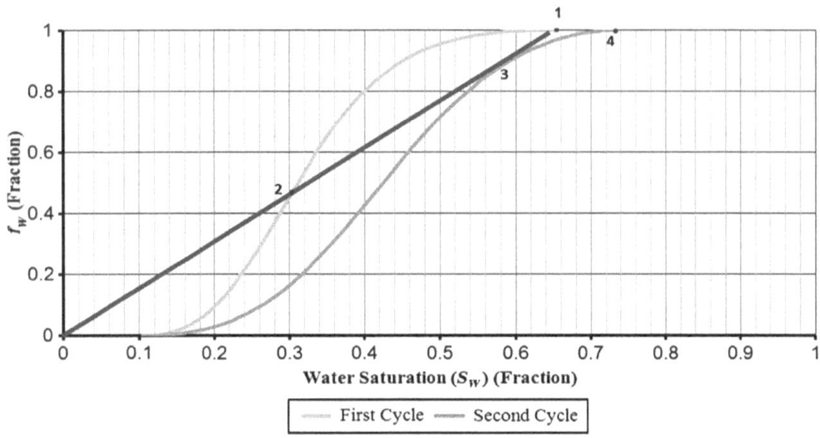

Figure 5.6 Fractional flow curves analysis for seawater (first) and low salinity water (second) injection cycles (Al-Shalabi et al., 2015c).

front. This definition physically makes more sense because it maintains the definition of mobility ratio as the mobility of the displacing fluid to the displaced fluid, and the chosen saturations physically exist. Fractional flow curves were analyzed to find the shock front and oil bank front saturations at each injection cycle, which is needed for calculating the corresponding relative mobility values. An example is shown in Fig. 5.6, the shock front saturation point is assigned number 3, and the oil bank front saturation is assigned number 2. The numbering shows the saturation profile where it starts at the producer condition (number 1) and ends at the injector condition (number 4). An example of the total relative mobility calculation at the oil bank front saturation is shown in Fig. 5.7.

Minimizing risk and uncertainty is a major part of any proposed IOR technique by highlighting the most significant design parameters and optimizing the entire process. Al-Shalabi et al. (2014e) optimized the LSWI process at field-scale for carbonate reservoirs by considering seven uncertain and decision parameters shown in Table 5.2.

5-spot LSWI pilot models were simulated using UTCHEM simulator with the empirical LSWI model. The design of experiment (DOE) method was used for sensitivity analysis and screening out the insignificant parameters (Fig. 5.8).

The response surface methodology (RSM) was implemented to optimize the LSWI cumulative oil recovery where a response surface was built (Fig. 5.9). In this study, the numerical optimization option in Design

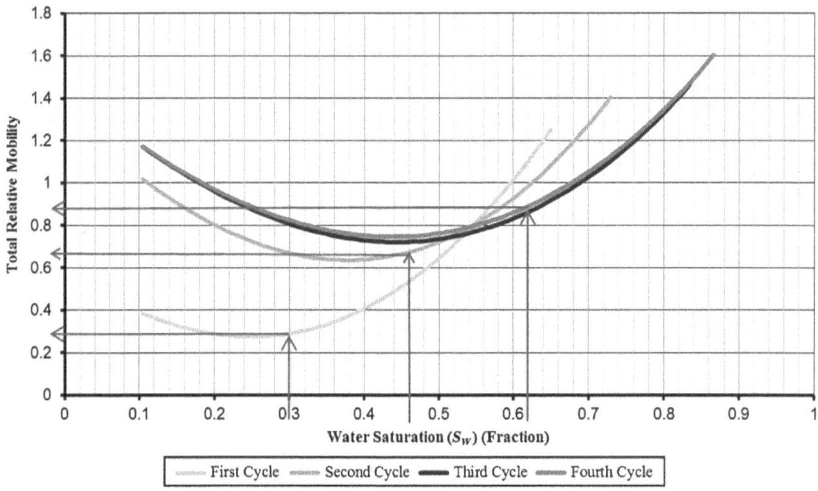

Figure 5.7 Total relative permeability calculations at the oil bank front saturation for different injection cycles (Al-Shalabi et al., 2015c).

Table 5.2 Two-level fractional factorial design parameters (Al-Shalabi et al., 2014e)

Variable			Coded symbol	Minimum value	Maximum value
Response variable		Maximizing cumulative oil recovery (%)	–	–	–
Process variables	**Uncertain variables**	Reservoir heterogeneity (V_{DP})	A	0.60	0.85
		kv/kh ratio	B	0.01	1.0
		S_{orw}	C	0.295	0.4
		S_{oi}	D	0.70	0.90
	Decision variables	Seawater slug size (PV)	E	1.0	3.0
		LSWI slug size (PV)	F	1.0	3.0
		Injected water salinity (meq/ml)	G	0.1	1.001

Expert software was utilized for the optimization process (Design Expert technical manual, 2011). The performed sensitivity analysis showed that three most important design parameters are LSWI slug size, reservoir heterogeneity characterized through Dykstra–Parson Coefficient (V_{DP}), and

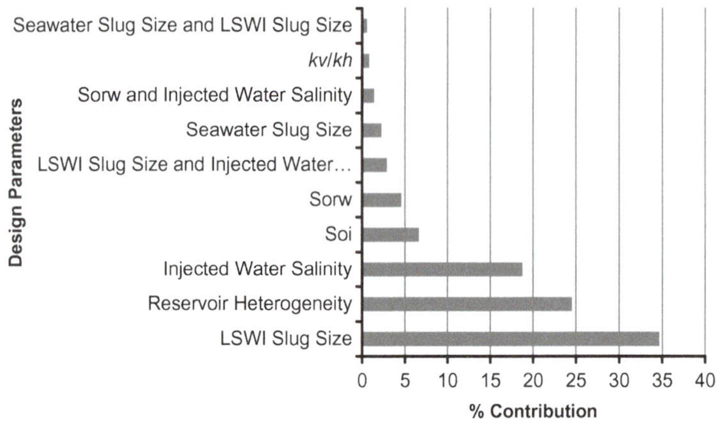

Figure 5.8 Effect of design parameters on cumulative oil recovery (Al-Shalabi et al., 2014e).

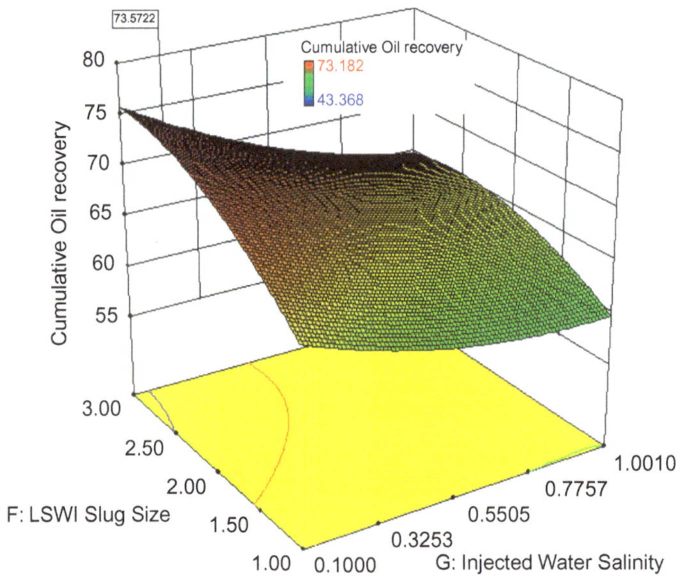

Figure 5.9 3D surface of cumulative oil recovery at varied values of LSWI slug size (PV) and injected water salinity (meq/ml) (Al-Shalabi et al., 2014e).

injected water salinity. Moreover, an optimized design was suggested and the results were validated against UTCHEM.

Moreover, Attar and Muggeridge (2015) evaluated the effect of heterogeneity of numerical simulation of the low salinity water injection using the models of both Jerauld et al. (2008) and Dang et al. (2013) along with vorticity heterogeneity index. They found that LSWI results in higher oil recovery compared with conventional seawater injection independent of the heterogeneity level. Also, they recommended an optimum slug size between 0.6 and 0.8 PV for heterogeneous cases due to dilution of the low salinity slug by cross flow between the high and low permeability zones. Usually, the low salinity is implemented in the field through injection of a finite length slug during both secondary and tertiary injection. This is because the produced water, which is usually more saline than the injected low salinity water, is typically reinjected.

Also, Dang et al. (2015) used their previously proposed LSWI model (Dang et al., 2013) to highlight the important role of clay distribution and content on oil recovery. The authors proposed an approach to model clay distribution and content. Moreover, they presented a new optimization concept based on well placement to take advantage of the clay distribution. The higher clay content in a certain zone leads to higher ion exchange, and hence, more wettability alteration toward a water-wet state, which is translated into higher oil recovery.

5.3 TRACER MODELING OF LSWI/EWI

Tracers were used to model low salinity water injection. Tracers are usually used for tracking the details of fluid movement, including the advancing fluid front, the breakthrough, the sweep efficiency, detection of flow barriers, and others. Modeling of tracer flow helps in interpreting complex field tracer tests, upon which advection-diffusion equations are used, and the diffusion term is usually neglected for most practical cases. Both active and passive traces as well as partitioning and nonpartitioning trances are included in the model. Passive tracers are the ones that do not change the physical properties of the phase in which they exist and are used mainly for tracking injected fluid. Nevertheless, active tracers are used for modeling physical and chemical effects not possible to model using black oil or compositional simulators, such as variation of water viscosity due to polymer, salinity effect on wettability, residual oil, relative permeability and capillary pressure, water PVT properties, mixing oils of different viscosity, and oil vaporization effect. Nonpartitioning tracer is the one that exists in a single phase; however, partitioning tracers partition between phases like single-well chemical tracer testing which the tracer partitions between oil and water phases (Verma et al., 2009).

Tracer tests are classified into interwell (IWTT) and single-well tests (SWTT). In the interwell tests, tracers are injected into one or more wells and produced from others. Nevertheless, in the single-well tracer test, tracers are injected into and then produced from the same well. These two tests have different advantages and disadvantages. For example, interwell tracer tests have access to larger areas of the investigated reservoirs, upon which more information can be gained; however, tracer recovery can take up to several years and the analysis can be very complex (Sharma et al., 2014). Among the information gained from interwell test are the general trends of fluid movements, layering, areal heterogeneity, directional flow trends, and saturations (Allison, 1988). On the other hand, the single-well tracer test takes fewer than 3 weeks to complete, and there is less uncertainty in the analysis due to the restriction to a small area of the investigated reservoir, which limits the amount of information gained from this test (Descant, 1989).

The single-well chemical tracer test (SWCTT) is also known as the residual oil saturation tracer test and the Exxon tracer test. This test uses chemical reactions involving different tracers. The primary objective of single-well tracer tests is determining fluid saturation. Other applications of this test include wettability-determining and permeability-determining tracer tests (Descant, 1989). Determination of residual oil saturation using SWCTT was first proposed by Tomich et al. (1973). This test uses the chromatographic separation of tracers with different partition coefficients, in conjunction with a tracer reaction. The theory and execution of the test were extensively covered by Deans and Majoros (1980), which includes the effects of fluid drift and other nonideal effects.

In SWCTT modeling, any number of tracers can be modeled, including water tracer, oil tracer, partitioning oil/water tracer, gas tracer, and partitioning gas/oil tracer. Reacting tracers are considered only for water/oil tracers. In the modeling of tracers, it is assumed that tracers neither occupy volume nor have any effect on physical properties. The overall tracer concentrations are computed from the species conservation equations, which include a reaction term for the reacting tracer. On the other hand, the tracer phase concentrations are calculated according to the tracer type: Water, oil, gas, or partitioning.

For the nonpartitioning tracers, the tracer phase concentration (C_{T_l}) is proportional to the ratio of the total tracer concentration (C_T) to the total concentration of water or oil depending on the tracer type (C_κ) as follows:

$$C_{T_l} = C_{kl}\frac{C_T}{C_\kappa}, \tag{5.15}$$

where T is the tracer type, including water or oil, l is the phase number, κ is the component number, and $C_{\kappa l}$ is the concentration of component κ in phase l.

On the other hand, for partitioning tracers, the partition coefficient for a water/oil tracer is defined based on water or oil pseudo-component concentration as the following:

$$K_T = \frac{C_{T2}}{C_{T1}}, \tag{5.16}$$

where C_{T1} and C_{T2} are the tracer concentrations in the water and oil pseudo-components, respectively. The tracer phase compositions are calculated from the tracer material balance equation as:

$$C_{Tl} = C_1 C_{T1} + C_2 C_{T2}, \tag{5.17}$$

By substituting Eq. (5.16) into Eq. (5.17), the resulting tracer phase concentrations are

$$C_{T1} = \frac{C_T}{C_1 + C_2 K_T}, \tag{5.18}$$

$$C_{T2} = K_T\frac{C_T}{C_1 + C_2 K_T}, \tag{5.19}$$

where C_1 and C_2 are the overall concentrations for water and oil species.

For tracer reaction, the hydrolysis of Ethyl Acetate to Ethanol is assumed to be of first order and irreversible as follows:

$$\underset{\text{Acetate}}{1CH_3COO[C_nH_{2n+1}]} + \underset{\text{Water}}{1H_2O} \rightarrow \underset{\text{Alcohol}}{1C_nH_{2n+1}[OH]} + \underset{\text{Acetic Acid,}}{1C_2H_4O_2} \tag{5.20}$$

where one mole of acetate (ETAC) generates one mole of product alcohol (ETOH). The reactions are modeled as:

$$\frac{\partial c_{ETAC}}{\partial t} = -kc_{ETAC}, \tag{5.21}$$

$$\frac{\partial c_{ETOH}}{\partial t} = kc_{ETOH}, \quad (5.22)$$

where k is the reaction constant in the unit of day^{-1}. Regional fluid drift is assumed to be negligible, which is a reasonable assumption based on a proper well selection where the effect of active producers and injectors in the nearby area is negligible. Moreover, the negligible effect of fluid drift is confirmed by checking the concentration profiles of material balance and cover tracers, which provides information about the shape and position of the reactive tracer (UTCHEM Technical Manual, 2000).

Verma et al. (2009) described modeling of improved oil recovery processes using tracers in an unstructured grid simulator. Also, modeling single-well chemical tracer tests is possible using this approach, which is used for determining the residual oil saturation before and after applying improved oil recovery methods. Negligible tracer can be used for modeling polymers and salinity effect on improving oil recovery. Decoupling of tracer equations from other component equations is the main advantage of using tracer for improved oil recovery modeling. Verma et al. (2009) modeled improving oil recovery by low salinity water injection using tracers that affect the rock and fluid properties. This was possible by modifying the relative permeability and capillary pressure curves, residual saturations along with water viscosity, density, and formation volume factor as function of salinity. A salt tracer was used in association with the water phase to model salinity variations. Properties of water at a given salinity were determined by interpolating between water properties at maximum and minimum salinity. The retention and adsorption of the amount of salt as a result of varying the salinity of injected water were modeled using the Langmuir adsorption isotherm.

Al-Shalabi et al. (2016) investigated modeling and simulation of SWCTT of low salinity water injection in a Middle Eastern carbonate reservoir using UTCHEM. Both Radial and Cartesian grid models were setup for a field-scale pilot. Two approaches were used to estimate remaining oil saturation to LSWI, including analytical and numerical methods. The plan used for the low salinity water injection single-well chemical tracer test (LSWI-SWCTT) is presented in Table 5.3, which includes a description of each step with the injection period. The LSWI-SWCTT is designed for 31 days, based on real field applications.

The results showed that both approaches give consistent values for remaining oil saturation for homogeneous Radial grid model. The Cartesian grid model was used to investigate the effect of heterogeneity on

Table 5.3 LSWI-SWCTT plan (Al-Shalabi et al., 2016)

Stage	Step	Description	Duration (days)
	1	Seawater injection	1.35
	2	Three tracers injection (NPA, EtAc, and IPA)	0.14
1	3	IPA tracer injection	0.55
	4	Well shut-in	4
	5	Well production	8.96
2	6	LSWI	4
	7	Seawater injection (restore initial salinity conditions)	1.35
	8	Three tracers injection (NPA, EtAc, and IPA)	0.14
	9	IPA tracer injection	0.55
	10	Well shut-in	4
	11	Well production	5.96
Total duration (days)			**31**

SWCTT, where a new numerical approach was proposed for estimating remaining oil saturation. The authors consider that the area in which oil saturation is below the original residual oil saturation is affected by the LSWI. Hence, the equation used to calculated the average residual oil saturation is

$$\overline{S}_{or} = \frac{\sum_i S_{o,i} V_i}{\sum_i V_i} \quad \forall i: S_{o,i} < 0.99 S_{or}^{HS}, \tag{5.23}$$

where $\overline{S}_{or}$ is the average residual oil saturation and V_i is the pore volume of gridblock i. The results validate the approach used and the implementation of both tracer reactions and LSWI model in UTCHEM.

The next chapter reviews the basics of geochemistry and its application in the LSWI/EWI field including geochemical and compositional modeling approach.

REFERENCES

Aladasani, A., Bai, B., Wu, U., 2012a. Investigating low salinity waterflooding recovery mechanisms in sandstone reservoirs. SPE Improved Oil Recovery Symposium, Tulsa, Oklahoma, USA, Paper SPE 152997.

Aladasani, A., Bai, B., Wu, Y., 2012b. Investigating low salinity waterflooding recovery mechanisms in carbonate reservoirs. SPE EOR Conference at Oil and Gas, West Asia, Muscat, Oman, Paper SPE 155560.

Alameri, W., Teklu, T.W., Graves, R.M., Kazemi, H., AlSumaiti, A.M., 2015. Experimental and numerical modeling of low salinity waterflood in a low permeability carbonate reservoir. SPE Western Regional Meeting, Garden Grove, California, USA, Paper SPE 174001.

Allison, S.B., 1988. Analysis and design of field tracers for reservoir description. Master's Thesis, The University of Texas at Austin, Austin, Texas, USA.

Al-Shalabi, E.W., 2014. Modeling the effect of injecting low salinity water on oil recovery from carbonate reservoirs. PhD Dissertation, The University of Texas at Austin, Texas, USA.

Al-Shalabi, E.W., Sepehrnoori, K., Delshad, M., 2014a. Mechanisms behind low salinity water injection in carbonate reservoirs. Fuel. 121, 11–19.

Al-Shalabi, E.W., Sepehrnoori, K., Pope, G.A., 2014b. Mysteries behind the low salinity water injection technique. J. Petrol. Eng. 2014, Article ID 304312. http://dx.doi.org/10.1155/2014/304312.

Al-Shalabi, E.W., Sepehrnoori, K., Delshad, M., Pope, G., 2014c. A novel method to model low-salinity water injection in carbonate oil reservoirs. SPE J. 20 (5), 1154–1166.

Al-Shalabi, E.W., Sepehrnoori, K., Pope, G., Mohanty, K., 2014d. A fundamental model for prediction oil recovery due to low salinity water injection in carbonate rocks. SPE Trinidad & Tobago Energy Resources Conference, Port of Spain, Trinidad and Tobago, Paper SPE 169911.

Al-Shalabi, E.W., Sepehrnoori, K., Delshad, M., 2014e. Optimization of the low salinity water injection process in carbonate reservoirs. SPE International Petroleum Technology Conference, Kuala Lumpur, Malaysia, Paper SPE 17821.

Al-Shalabi, E.W., Sepehrnoori, K., Delshad, M., 2015a. Simulation of wettability alteration by low salinity water injection in water-flooded carbonate cores. J. Petrol. Sci. Technol. 33 (5), 604–613.

Al-Shalabi, E.W., Sepehrnoori, K., Delshad, M., 2015b. Numerical simulation of the LSWI effect on hydrocarbon recovery from carbonate rocks. J. Petrol. Sci. Technol. 33 (5), 595–603.

Al-Shalabi, E.W., Sepehrnoori, K., Pope, G., 2015c. New mobility ratio definition for estimating volumetric sweep efficiency of low salinity water injection. Fuel. 158, 664–671.

Al-Shalabi, E.W., Luo, H., Delshad, M., Sepehrnoori, K., 2016. Single-well chemical tracer modeling of low salinity water injection in carbonates. SPE Reservoir Evaluation and Engineering Journal, In press.

Attar, A., Muggeridge, A., 2015. Impact of geological heterogeneity on performance of secondary and tertiary low salinity water injection. SPE Middle East Oil & Gas Show and Conference, Manama, Bahrain, Paper SPE 172775.

Chandrasekhar, S., Mohanty, K.K., 2013. Wettability alteration with brine composition in high temperature carbonate reservoirs. SPE Annual Technical Conference and Exhibition, New Orleans, Louisiana, USA, Paper SPE 166280.

Craig Jr., F.F., 1993. The Reservoir Engineering Aspects of Waterflooding, In: SPE of AIME, Dallas.

Dang, C.T.Q., Nghiem, L.X., Chen, Z., Nguyen, Q.P., 2013. Modeling low salinity waterflooding: ion exchange, geochemistry and wettability alteration. SPE Annual Technical Conference and Exhibition, New Orleans, Louisiana, USA, Paper SPE 166447.

Dang, C., Nghiem, L., Nguyen, N., Chen, Z., Nguyen, Q., 2015. Modeling and optimization of low salinity waterflood. SPE Reservoir Simulation Symposium, Houston, Texas, USA, Paper SPE 173194.

Deans, H.A., Majoros, S., 1980. The single-well chemical tracer method for measuring residual oil saturation. U.S. DOE Report BC/20006-18.

Descant, F.J., 1989. Simulation of single-well tracer flow. Master's Thesis, The University of Texas at Austin, Austin, Texas, USA.

Design-Expert Software, 2011. Technical Manual, version 8.

Jerauld, G.R., Lin, C.Y., Webb, K.J., Seccombe, J.C., 2008. Modeling low salinity waterflooding. SPE Reserv. Eval. Eng. 11 (6), 1000–1012.

Jin, M., 1995. A study of non-aqueous phase liquid characterization and surfactant remediation. PhD Dissertation, The University of Texas at Austin, Texas, USA.

Lemon, P., Zeinijahromi, A., Bedrikovestsky, P., Shahin, I., 2011. Effects of injected water salinity on waterflood sweep efficiency through induced fines migration. J. Can. Petrol. Technol. 50 (9–10), 82–94.

Pope, G.A., Wu, W., Narayanaswamy, G., Delshad, M., Sharma, M.M., Wang, P., 2000. Modeling relative permeability effects in gas-condensate reservoirs with a new trapping model. SPE Reserv. Eval. Eng. 3 (2), 171–178.

Sharma, A., Shook, G.M., Pope, G.A., 2014. Rapid analysis of tracers for use in eor flood optimization. SPE Improved Oil Recovery Symposium, Tulsa, Oklahoma, USA, Paper SPE 169109.

Stumm, W., Morgan, J., 1996. Aquatic Chemistry. John Wiley & Sons, Inc, New York.

Tomich, J.F., Dalton, R.L., Deans, H.A., Shallenberger, L.K., 1973. Single-well tracer method to measure residual oil saturation. J. Petrol. Technol. 255, 211–218.

Tripathi, I., Mohanty, K.K., 2008. Instability due to wettability alteration in displacements through porous media. Chem. Eng. Sci. 63 (21), 5366–5374.

UTCHEM – 9.0 Technical Documentation, 2000. The University of Texas at Austin, Volume II, Texas, USA.

Verma, S., Adibhatla, B., Leahy-Dios, A., Willingham, T., 2009. Modeling improved recovery methods in an unstructured grid simulator. International Petroleum Technology Conference, Doha, Qatar, Paper SPE 13920.

Wu, Y., Bai, B., 2009. Efficient simulation for low salinity waterflooding in porous and fractured reservoirs. SPE Reservoir Simulation Symposium, The Woodlands, Texas, USA, Paper SPE 118830.

Yousef, A.A., Al-Saleh, S., Al-Kaabi, A., Al-Jawfi, M., 2011. Laboratory investigation of the impact of injection-water salinity and ionic content on oil recovery from carbonate reservoirs. SPE Reserv. Eval. Eng. 14 (5), 578–593.

Yousef, A.A., Al Saleh, S., Al Jawfi, M., 2012. Improved/enhanced oil recovery from carbonate reservoirs by tuning injection water salinity and ionic content. SPE Improved Oil Recovery Symposium, Tulsa, Oklahoma, USA, Paper SPE 154076.

Yu, L., Evje, S., Kleppe, H., Karstad, T., Fjelde, I., Skjaeveland, S.A., 2009. Spontaneous imbibition of seawater into preferentially oil-wet chalk cores- Experiments and Simulations. J. Petrol. Sci. Eng. 66 (3–4), 171–179.

CHAPTER SIX

Geochemistry of Low Salinity/ Engineered Water Injection Processes

Contents

This chapter includes a review of basic geochemistry modeling and a description of different geochemical reactions including aqueous, dissolution/precipitation, multiion exchange, and surface complexation reactions, as well as the application of geochemistry in the field of low salinity/engineered water injection.

6.1 BASIC GEOCHEMISTRY MODELING

Subsurface geochemistry and transport of chemicals have essential impact upon the fate of different improved oil recovery (IOR) processes. For example, an optimal salinity gradient design is the key for a successful surfactant-polymer (SP) flooding, in which minimum interfacial tension is achieved in the focused emulsification region (Nelson and Pope, 1978;

Low Salinity and Engineered Water Injection for Sandstone and Carbonate Reservoirs.
DOI: http://dx.doi.org/10.1016/B978-0-12-813604-1.00006-7

Lake, 1989; Delshad et al., 1996; Green and Willhite, 1998). Moreover, the increase in pH significantly reduces the adsorption of surfactant; hence, it is preferable to combine alkali to the SP flooding (Nelson et al., 1984; Martin et al., 1985; Sheng, 2013). Alkali may also react with the acid in crude oil and generate in-situ soap, which strengthens the emulsification capability within the reservoir. This synergy needs to be specifically taken care of because the optimal salinity of the surfactant-soap mixture may vary significantly based on their chemical attributes and proportions. Another example is low salinity/engineered water injection (LSWI/EWI) upon which the amounts of certain geochemical species, such as sulfate ion, calcium ion, magnesium ion, anhydrite, and calcite solid species, have pronounced effects in changing the rock wettability and flow functions (Lager et al., 2008; Austad et al., 2010; Al-Shalabi et al., 2014 and 2015).

The previous studies strongly indicate the need for a robust geochemical simulator, which accurately predicts the concentrations of geochemical species and hence oil recovery for different IOR techniques.

6.1.1 Chemical thermodynamics of equilibrium process

In thermodynamics, multiphase, multicomponent systems are in thermodynamic equilibrium when Gibbs free energy is a minimum at constant temperature (T), pressure (P), and composition (N). There are other equivalent conditions for equilibrium, but this condition is the most convenient to apply. Under thermodynamic equilibrium conditions, there is no change in the state of the system. The mathematical expression of thermodynamic equilibrium is as follows:

$$\mathrm{d}G^t = 0, \tag{6.1}$$

where G^t is the total Gibbs free energy and $\mathrm{d}G^t$ is the differential variation of the total Gibbs free energy at constant T, P, and N. For L phases and m species, the total Gibbs free energy is defined as

$$G^t = \sum_{l=1}^{L} \sum_{i=1}^{m} n_i^l \overline{G}_i^l, \tag{6.2}$$

where n_i^l is the moles of species i in phase l and $\overline{G}_i^l$ is the partial molar Gibbs free energy of species i in phase l. $\overline{G}_i^l$ is often called the chemical potential and denoted by μ_i^l. When the composition of the system has changed due to a certain chemical reaction (*r*), the condition for reaching thermodynamic equilibrium is

$$\Delta G_r = 0, \tag{6.3}$$

where ΔG_r is the change in Gibbs free energy for the reaction (*r*) and is defined as

$$\Delta G_r = \sum_{i=1}^{m} \nu_{ri}\mu_i, \tag{6.4}$$

and the chemical potential of species *i* is given by

$$\mu_i = G_i^{\mathrm{o}} + RT \ln a_i, \tag{6.5}$$

where a_i is the activity of the chemical species *i*, G_i^{o} is the standard state Gibbs free energy of formation of the species *i*, and ν_{ri} is the stoichiometric coefficient of reaction (*r*) with species *i*. The standard state is referred to as the pure state of the species *i* at the temperature of the solution and one atmospheric pressure. Equations (6.3–6.5) result in the following equation:

$$\ln \prod_{i=1}^{m} a_i^{\nu_{ri}} = \frac{-\sum_{i=1}^{m} \nu_{ri} G_i^{\mathrm{o}}}{RT}. \tag{6.6}$$

The right side of Equation (6.6) is a function of temperature only; however, the left side is a function of the activities of the reactants and products under equilibrium conditions. The equilibrium constant of the reaction (*r*) at a certain temperature (*T*) is defined as

$$K_r(T) = \exp \frac{-\sum_{i=1}^{m} \nu_{ri} G_i^{\mathrm{o}}}{RT}. \tag{6.7}$$

Hence, Equation (6.6) can be written as

$$K_r(T) = \prod_{i=1}^{m} a_i^{\nu_{ri}}. \tag{6.8}$$

The calculation of the equilibrium constant (*K*) from Equation (6.7) is easy for standard temperature of 298.15 K because the standard state free energies of formation of chemical species are tabulated at this temperature. For temperatures other than 298.15 K, the equilibrium constant can be computed using the van't Hoff equation (Van't Hoff, 1884):

$$\frac{\mathrm{d}[\ln K_r(T)]}{\mathrm{d}T} = \frac{\Delta H_r^{\mathrm{o}}(T)}{RT^2}, \tag{6.9}$$

where ΔH_r° is the standard heat of reaction for the reaction (*r*) and is a function of temperature; however, in many cases, ΔH_r° can be assumed over a narrow range of temperature to be approximately constant and the integration of Equation (6.9) results in

$$\ln\frac{K_r(T_2)}{K_r(T_1)} = \frac{\Delta H_r^\circ}{R}\left[\frac{1}{T_1} - \frac{1}{T_2}\right]. \tag{6.10}$$

Equation (6.10) can be used for calculating the equilibrium constant of a reaction at any temperature T_2 from the equilibrium constant value at temperature T_1 and the standard heat of reaction (Sandler, 2006). An analytical expression of Equation (6.10) is presented below (Parkhurst and Appelo, 2013):

$$\log_{10} K = A_1 + A_2 T + \frac{A_3}{T} + A_4 \log_{10} T + \frac{A_5}{T^2}, \tag{6.11}$$

where A_1–A_5 are constants and T is the temperature in K. The equilibrium reaction constants for species and solubilities of minerals (K_P) can also be defined as functions of pressure as follows:

$$\log K_P = \log K_{P=1} - \frac{\Delta V_r}{2.303RT}(P-1), \tag{6.12}$$

where P is the pressure in atm, T is the temperature in K, ΔV_r is the volume change of the reaction (cm^3/mol), and R is the gas constant (82.06 atm cm^3/(mol K)). This dependency is defined using molar volume of the solids with the volumes of the solute species (Parkhurst and Appelo, 2013).

The thermodynamic state of a mineral relative to a solution is defined by saturation index (SI) which is expressed as follows:

$$\mathrm{SI} = \log 10\left(\frac{\mathrm{IAP}}{K}\right), \tag{6.13}$$

where IAP is the ion activity product and K is the equilibrium constant. If SI is less than zero and the mineral is present, the mineral could dissolve, but not precipitate. If SI is greater than zero, the mineral could precipitate, but not dissolve. When SI is equal to zero, this indicates that the minerals are in equilibrium with the solution (Parkhurst and Appelo, 2013).

6.1.2 Activity coefficient models

The activities of chemical species in solution are related to their molar concentration through their individual activity coefficients (γ_i) as follows:

$$a_i = \gamma_i C_i. \tag{6.14}$$

For ideal solutions, the activity coefficients of the species are equal to unity; hence, the activities in Equation (6.8) can be replaced by molar concentrations. For nonideal solution, the activity coefficient is calculated using the activity coefficient models. These models describe the relation between the activity coefficients of species and ionic strength of the solution. The ionic strength of the aqueous solution (I) was previously defined in Equation (5.8), Chapter 5. Davies equation and extended or WATEQ, a computer program for calculating chemical equilibria (Truesdell and Jones, 1974), Debye–Huckel equation are two of the most commonly used activity coefficient models for charged species; however, for the uncharged species, Setchenow equation is usually used (Sandler, 2006).

- *Davies Equation*

$$\log \gamma_i = -Az_i^2\left(\frac{\sqrt{I}}{1+\sqrt{I}} - 0.3I\right), \tag{6.15}$$

- *Extended or WATEQ Debye–Huckel Equation*

$$\log \gamma_i = -\frac{Az_i^2\sqrt{I}}{1+Ba_i^{\mathrm{o}}\sqrt{I}} + b_i I, \tag{6.16}$$

- *Setchenow Equation*

$$\log \gamma_i = b_i I. \tag{6.17}$$

Equation (6.16) is the WATEQ Debye–Huckel equation and when b_i is zero, it becomes the Extended Debye–Huckel equation. For the extended Debye–Huckel equation, a_i^{o} is the ion size parameter in Angstrom, whereas for WATEQ Debye–Huckel equation, a_i^{o} and b_i are ion specific parameters fitted from mean-salt activity coefficient. Equation (6.16) becomes the Setchenow equation (Equation (6.17)) when the first term of activity coefficient is zero, and usually b_i is assumed to be 0.1 for all uncharged species (Parkhurst and Appelo, 2013). In Equations (6.15) and (6.16), A and B are temperature-dependent constants, which have the following forms (Manov et al., 1943):

$$A = 1.8246 \times 10^6(\varepsilon_w T)^{-3/2}\ (\mathrm{mol}^{-1/2}\,\mathrm{L}^{1/2}), \tag{6.18}$$

$$B = 50.29(\varepsilon_w T)^{-1/2}\ (\mathrm{mol}^{-1/2}\,\mathrm{L}^{1/2}\,\mathrm{\mathring{A}}), \tag{6.19}$$

where T is the absolute temperature in Kelvin, and ε_w is the dielectric constant of water as a function of temperature (Malmberg and Maryott, 1956) expressed as

$$\begin{aligned}\varepsilon_w &= 87.74 - 0.4008(T - 273.15) + 9.398 \times 10^{-4}(T-273.15)^2 \\ &\quad - 1.410 \times 10^{-6}(T-273.15)^3.\end{aligned} \tag{6.20}$$

There are more accurate activity coefficient models for high-salinity solutions, such as Pitzer model (Pitzer, 1991) and specific ion interaction theory model (Grenthe et al., 1997). Taking Pitzer model as an example, it combines the Debye–Huckel equation with additional terms in the form of a virial equation, which are intended to account for short-range interactions among ions (Pitzer, 1991). The general form of Pitzer model is as follows:

$$\ln(\gamma_i) = \ln(\gamma_i^{\mathrm{dh}}) + \sum_j D_{ij}(I)m_j + \sum_j \sum_k E_{ijk} m_j m_k, \tag{6.21}$$

where γ_i^{dh} refers to a Debye–Huckel term, and D_{ij} and E_{ijk} are second and third virial coefficients, defined for each pair and triplet of ions in solution.

For pure solids, the activities of the species are considered equal to unity. The water activity equation is given by

$$a_{\mathrm{H_2O}} = 1 - 0.017 \sum_{i=1}^{N_{\mathrm{aq}}} m_i, \tag{6.22}$$

where i is an aqueous specie and N_{aq} is the total number of aqueous species. The equation is an approximation of Raoult's law as was reported by Garrels and Christ (1965).

6.1.3 Basic geochemistry reactions

A general form of geochemical reaction equilibrium can be represented by

$$aP + bQ \leftrightarrow cT, \tag{6.23}$$

where P, Q, and T are the reactive geochemical species, and a, b, and c, are their reaction stoichiometric coefficients, respectively. The reaction equilibrium constant (K) can be written as

$$K = \frac{a_T^c}{a_P^a \cdot a_Q^b}, \tag{6.24}$$

where a_P, a_Q, and a_T are activities of geochemical species P, Q, and T, respectively.

Aqueous electrolytes chemistry, dissolution/precipitation of minerals, ion-exchange reactions with the matrix and micelle, and the reaction of acidic components of the oil with bases in the aqueous solution are the reactions included in the geochemical models. Each of these reactions is discussed below:

- *Aqueous (Homogeneous) Reactions*

  In these reactions, the free ions (e.g., Na^+, K^+, Mg^{2+}, Ca^{2+}, Cl^-, SO_4^{2-}, etc.) react to form complex ions (e.g., $NaCl_{(aq)}$, $CaSO_{4(aq)}$, $MgSO_{4(aq)}$, etc.). The reactions are homogeneous because all components involved in the reactions exist in one phase. Usually, a common practice in geochemistry is to assume all aqueous reactions are fast and hence at equilibrium. An example of aqueous (homogeneous) reactions is given by

$$Na^+ + Cl^- \leftrightarrow NaCl_{(aq)}. \tag{6.25}$$

- *Dissolution/Precipitation (Heterogeneous) Reactions*

  These relations are also called mineral reactions or solubility product reactions because in these reactions, the equilibrium constant (K) is called the solubility product (K_{sp}). The reactions are heterogeneous because the components involved in these reactions are in different phases. An example of mineral reactions is calcite dissolution/precipitation, which is defined as

$$Ca^{2+} + CO_3^{2-} \leftrightarrow CaCO_{3(s)}. \tag{6.26}$$

  Mineral reactions are usually considered as kinetically controlled reactions unless the reaction is fast then local equilibrium assumption could be valid.

- *Kinetic vs Local Equilibrium Reactions*

  Kinetic reactions describe the rate (or how fast) a reaction takes place. These reactions are usually described in comparison to the local equilibrium reactions. Usually, homogeneous (aqueous reactions) are considered at local equilibrium; however, heterogeneous (dissolution/precipitation of the solids) reactions are better represented by kinetics. Mineral reactions could be assumed at local equilibrium if the reactions are fast (Zhu and Anderson, 2002). The dissolution or precipitation rate of a mineral is calculated as follows:

$$r_\beta = \hat{A}_\beta k_\beta \left(1 - \frac{Q_\beta}{K_\beta}\right), \tag{6.27}$$

where $\hat{A}_\beta$ is the reactive surface area of the reactant mineral β per unit bulk volume of porous medium (m^2/m^3), k_β is the rate constant of mineral reaction β ($mol/m^2\,s$), K_β is the equilibrium constant of mineral dissolution/precipitation reaction, Q_β is the activity product of mineral β dissolution/precipitation reaction, and r_β is the dissolution/precipitation rate per unit bulk volume of porous medium (mol/(m^3 s)). The reaction rate constant (k_β) is temperature dependent as follows:

$$k_\beta = k_{0,\beta} \exp\left[-\frac{E_a}{R}\left(\frac{1}{T} - \frac{1}{T_0}\right)\right], \tag{6.28}$$

where E_a is the activation energy for reaction (J/mol), $k_{0,\beta}$ is the reaction-rate constant at reference temperature T_0, R is the universal gas constant (8.314 J/(mol K)), T is the temperature (K), and T_0 is the reference temperature (K).

- *Exchange Reactions*

  These are reactions between geochemical species in the solution and an electrolyte attached to a solid surface. The general form of these reactions can be written as follows:

$$X^-A^+ + B^+ \leftrightarrow X^-B^+ + A^+, \tag{6.29}$$

where X^- is the exchanger, and both A and B are exchange species. Clays are one of the most important exchangers from a geochemistry point of view. In sandstone reservoirs, especially the sandstone reservoirs, there could be an insoluble negative-charged exchanger (X) within the clay, which adsorbs a fixed amount of equivalence of cations per pore volume, cation exchange capacity (CEC). These adsorbed ions could exchange with the cations in the aqueous phase. We regard the exchange reaction equation as one exchange species replacing another exchange species, because the relative exchange equilibrium constant between two exchange species is easier to obtain from the experiment or the literature (Farajzadeh et al., 2012). An example of exchange reactions in the presence of clay is as follows:

$$Na^+ + \frac{1}{2}(Ca - X_2) \leftrightarrow (Na - X) + \frac{1}{2}Ca^{2+}. \tag{6.30}$$

The equilibrium constant for exchange reactions is called selectivity coefficient (K') that is expressed for the example above as

$$K'_{\mathrm{Na/Ca}} = \frac{\zeta(\mathrm{Na-X})\left[\mathrm{Ca^{2+}}\right]^{0.5}}{(\zeta(\mathrm{Ca-X_2}))^{0.5}[\mathrm{Na^+}]} \times \frac{\gamma^{0.5}_{\mathrm{Ca^{2+}}}}{\gamma_{\mathrm{Na^+}}}, \quad (6.31)$$

where $\zeta(\mathrm{Na-X})$ and $\zeta(\mathrm{Ca-X_2})$ are the equivalent fractions of Na^+ and Ca^{2+} on the exchanger, respectively.

In carbonate reservoirs, the positive-charged exchanger (X) on the carbonate surface adsorbs a fixed amount of equivalence of anions per pore volume, anion exchange capacity. Hence, an anion exchange might occur on the positively charged carbonate surface, where the sulfate anion in water exchanges with the negatively charged carboxylic acid group in oil. This exchange reaction results in releasing oil ganglia and altering the rock wettability toward a more water-wet state. With increasing the reservoir temperature, the sulfate adsorption on rock surface increases and the effect of this reaction becomes more pronounced (Zhang et al., 2006). This anionic exchange reaction can described as follows:

$$\mathrm{SO_4^{2-} + 2CH_3COO-X = 2CH_3COO^- + SO_4 - X_2}, \quad (6.32)$$

where X represents the carbonate rock and CH_3COO^- represents the carboxylic acid group of the oil. The selectivity coefficient (K') for this reaction can be expressed as the following:

$$K'_{\mathrm{SO_4/CH_3COO}} = \frac{\zeta((\mathrm{SO_4-X_2})[\mathrm{CH_3COO^-}]^2}{(\zeta(\mathrm{CH_3COO-X}))^2\left[\mathrm{SO_4^{2-}}\right]} \times \frac{\gamma^2_{\mathrm{CH_3COO^-}}}{\gamma_{\mathrm{SO_4^{2-}}}}, \quad (6.33)$$

where $\zeta(\mathrm{SO_4-X_2})$ and $\zeta(\mathrm{CH_3COO-X})$ are the equivalent fractions of SO_4^{2-} and CH_3COO^- on the exchanger, respectively.

Different from the aqueous species reactions and solid dissolution/precipitations, the activities of the exchange species are not calculated from concentration, but from the equivalent fractions of the exchange species. The equivalent fraction of the ith exchange species ζ_i is calculated by

$$\zeta_i = \frac{z_i c_i}{\sum_{i=1}^{N_{ex}} z_i c_i} = \frac{z_i c_i}{\mathrm{CEC}} = \frac{z_i c_i}{\mathrm{AEC}}, \quad (6.34)$$

where c_i is the concentration of the exchange species, z_i is the valence number of the exchange species, and N_{ex} is the total number of exchange species.

The use of selectivity coefficients follows the Gaines–Thomas convention (Gaines and Thomas, 1953). Moreover, it should be noted that selectivity coefficients are operational variables and not thermodynamic variables like the equilibrium constants. It is worth mentioning that exchange reactions are similar to the adsorption phenomena (Green and Willhite, 1998).

- *Surfactant-Associated Exchange Species Reactions*

  If anionic surfactant micelles and/or in-situ soap are present, aqueous cation ions can also exchange with the surfactant-associated cation ions, which is similar to the exchange reactions described above. Hirasaki (1982) pointed out that reaction equilibrium constant (K) in this case is not constant, but a function of total anionic surfactant concentration:

$$K_i = \beta_i(c_{\text{surf}} + c_{\text{A}^-}), \tag{6.35}$$

  where β_i is the coefficient of the ith surfactant-associated exchange equilibrium constant, and c_{surf} and c_{A^-} are the concentrations (molarities) of surfactant and oleic acid anion, respectively. The activities of the surfactant-associated exchange species are calculated from the equivalent fractions, which is similar to the general exchange reactions.

- *Oleic Acid Species Reactions*

  If the oleic acid reaction is considered in the calculation, then one more element, A, one more independent aqueous species, HA_o, and two additional dependent aqueous species, HA_w and A^-, need to be added to the geochemical system. The partition of acidic/basic components between the aqueous and oleic phase is given by Equation (6.34) is shown in Fig. 6.1.

$$HA_w = HA_o. \tag{6.36}$$

The equilibrium constant controlling the partition reaction is given by

$$K_{wo} = \frac{[HA_w]}{[HA_o]}, \tag{6.37}$$

where K_{wo} is the partition coefficient, $[HA_w]$ and $[HA_o]$ are the concentrations of the acidic component in the aqueous and oleic phases, respectively.

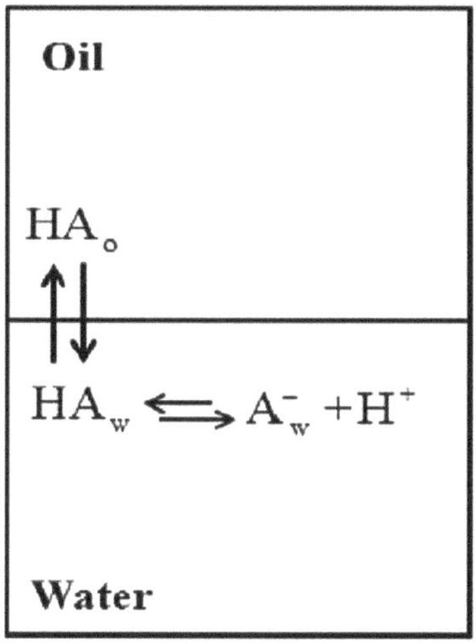

Figure 6.1 Equilibrium in water/oil/naphthenic acid systems at low pH (Havre et al., 2003).

Further, the partitioned ion in the aqueous phase dissociates in the aqueous phase as follows:

$$HA_w = A_w^- + H^+, \tag{6.38}$$

with the following a reaction equilibrium constant:

$$K_a = \frac{[H^+][A_w^-]}{[HA_w]}. \tag{6.39}$$

- *Surface Complexation Reactions*

  Surface complexation reactions are considered as a general form of the ion-exchange reactions except the fact that species are attached to the existing functional groups of the solid surface of amorphous aluminosilicates, metal oxides/hydroxides, and organic matters. These are reactions in which the carboxylic–divalent (organometallic) complexes on the rock surface are exchanged with divalents with no attached carboxylic branch and/or monovalent ions (Lager et al., 2011). In addition, in the surface complexation reactions, the electrical potential of the surface is considered in the sorption of the species existing on the surface.

  Surface complexation models describe surface reactivity behavior. There are two surface complexation models to describe the surface reactions. These models are the electrical diffusive layer (EDL) and the nonelectrical diffusive layer (NEDL) model. The NEDL model is used when surface potential effects are not significant. On the other hand, when surface potential dominates surface complexation, the EDL model is used. In the EDL model, the activity of aqueous species is corrected through a Boltzmann factor. Hence, the equilibrium constant for such reactions becomes as follows:

$$K_{\text{solution}} = K_{\text{surface}} \cdot \exp\left(-\frac{\Delta z F \Psi}{RT}\right), \tag{6.40}$$

where K_{solution} is the equilibrium constant of the aqueous solution which is mostly used in modeling programs, K_{surface} is the surface equilibrium constant obtained the literature (e.g., Dzombak and Morel, 1990), Δz is the net charge of surface species (sites), F is Faraday constant, Ψ is surface potential, R is the universal gas constant, and T is the temperature. The surface potential (Ψ) is calculated using the Goy–Chapman model:

$$\sigma = \left(8RT\varepsilon\varepsilon_0 I \times 10^3\right)^{1/2} \cdot \sinh\left(\frac{zF\Psi}{2RT}\right), \tag{6.41}$$

where σ is the surface charge density, ε is the dielectric constant of water, I is the ionic strength of solution, z is the ion charge, and ε_0 is the permittivity of free space. It should be noted that the surface charge density (σ) can be calculated from the concentration of solid surface species (C/m^2).

Moreover, there are other more complex version of surface complexation models, such as the triple layer model in which the surface layer can be made up of two different layers: one closely bound to the surface and one less closely bound. More information about this model can be found elsewhere (Davis et al., 1978).

An example of surface complexation reaction is as follows:

$$> CaOH_2^+ + SO_4^{2-} \leftrightarrow > CaSO_4^- + H_2O, \tag{6.42}$$

where $> CaOH_2^+$ and $> CaSO_4^-$ are surface species and represented by the symbol "$>$".

6.2 MECHANISTIC MODELING OF LSWI/EWI

The mechanistic modeling of LSWI/EWI is based on the involved geochemical reactions. Most of the researchers in the literature believe in wettability alteration as the underlying mechanism behind incremental oil recovery by LSWI/EWI. Therefore, the aqueous–rock geochemistry should be linked to the change in the wettability state of the rock toward a more water-wet system.

The geochemistry engine can also be linked to a compositional reservoir simulator to include the buffering effect of commonly occurring hydrocarbon phase components (e.g., CO_2, CH_4, and acidic/basic) on the aqueous–rock geochemistry. For this purpose, the reactive transport processes should be linked to the phase equilibrium calculations in an EOS compositional model.

Korrani (2014) presented the coupling of a compositional reservoir simulator, UTCOMP, to a geochemistry engine, IPHREEQC (Charlton and Parkhurst, 2011). The mass conservation equation was first implemented in UTCOMP in order to model the transport of different geochemical elements (e.g., Na, Ca, Mg, S, C, etc.). However, these geochemical species are treated as active tracers, and hence IPHREEQC was coupled in the simulator to determine the equilibrium state of different geochemical species in the reservoir gridblocks at each time step. A description of UTCOMP simulator, PHREEQC geochemistry engine, and the coupling is described here.

6.2.1 Description of the UTCOMP simulator

UTCOMP, developed at the University of Texas at Austin, is a non-isothermal, three-dimensional, equation-of-state (EOS), IMPEC (implicit pressure and explicit phase saturations and compositions), compositional reservoir simulator used for simulating different enhanced oil recovery processes including miscible and immiscible gas flooding (UTCOMP Technical Documentation, 2003).

The UTCOMP simulator is based on Acs et al. (1985) as a volume-balance approach with some other modifications. UTCOMP solution is IMPEC where the pressure at each gridblock is solved implicitly and the component mole rather than phase saturation is calculated explicitly. UTCOMP is capable of modeling four-phase flow behaviors that are an aqueous phase, an oleic phase, a gaseous phase, and a second nonaqueous liquid phase. The aqueous phase is entirely water and hydrocarbon components can be soluble in the aqueous phase. The hydrocarbon phase behavior is modeled using both the Peng Robinson (PR) EOS (Peng and

Robinson, 1976) and a modified version of the Redlich Kwong (RK) EOS (Turek et al., 1984). Based on Jhaveri and Youngren (1988) work, a volume-shift parameter option was designed to adjust the hydrocarbon phase density calculations. A detailed description of the UTCOMP simulator is available elsewhere (Chang, 1990).

The currently available features of UTCOMP are summarized as the following:

- Rigorous and simplified flash calculations (including four-phase flash-calculation capability)
- *K*-value method for phase-behavior calculations
- Higher order total variation diminishing finite-difference method
- Full physical-dispersion tensor
- Variable-width cross-section option
- Vertical or horizontal well capability
- Tracer-flood capability
- Polymer-flood capability
- Dilute-surfactant option with both equilibrium and nonequilibrium mass transfer
- Gas–foam-flood capability (P_c^* model and table-look-up approach)
- Black-oil model
- Asphaltene precipitation model
- CO_2 sequestration in aquifers

A third-order finite difference method is used by UTCOMP to reduce numerical dispersion and for grid orientation control. Physical dispersion is simulated using the full dispersion tensor, and the elements of the dispersion tensor emerge from molecular diffusion and mechanical dispersion. Relative permeability, interfacial tension, and capillary pressure are included. Relative permeability and capillary pressure can be dependent on interfacial tension through the concept of capillary number. MacLeod–Sugden correlation is used to calculate the interfacial tension between hydrocarbon phases (Reid et al., 1987). Hydrocarbon viscosity is calculated using the Lohrenz et al. (1964) correlation while the water viscosity is constant. It is worth mentioning that UTCOMP has been recently enhanced with Embedded Discrete Fracture Model to handle complex fracture geometries (Shakiba, 2014).

Four predominant forces, viscous, gravity, capillary, and dispersion forces, cause the transport of various chemical species in the multicomponent, multiphase flow porous media. The component mole balance equation is described as

$$\frac{\partial N_i}{\partial t} - V_b \vec{\nabla} \cdot \left[\sum_{j=1}^{n_p} \xi_j \lambda_j x_{ij} \left(\nabla P_j - \gamma_j \nabla D \right) + \varphi \xi_j S_j \vec{\vec{K}}_{ij} \nabla x_{ij} \right] - q_i \\ = 0 \quad \text{for } i = 1, 2, \ldots, n_c, \tag{6.43}$$

where the equation is written in terms of moles per unit time, N_i is the number of moles for component i, V_b is the bulk volume of a gridblock, ξ_j is the molar density of phase j, x_{ij} is the mole fraction of component i in phase j, γ_j is the specific weight of phase j, D is depth, and q_i is the molar injection (positive) or production (negative) rate for component i. The mobility for phase j is defined as follows:

$$\lambda_j = k \frac{k_{rj}}{\mu_j}. \tag{6.44}$$

Physical dispersion is modeled using the full dispersion tensor as below:

$$\vec{\vec{K}}_{ij} = \begin{bmatrix} K_{xx} & K_{xy} & K_{xz} \\ K_{yx} & K_{yy} & K_{yz} \\ K_{zx} & K_{zy} & K_{zz} \end{bmatrix}_{ij}. \tag{6.45}$$

The number, amounts, and composition of all equilibrium phases are determined using the phase-equilibrium calculations. The equilibrium solution must meet three conditions; the molar balance constraint must be preserved, the chemical potentials for each component must be the same in all phases, and the Gibbs free energy at constant temperature and pressure must be a minimum. The first partial derivative of the total Gibbs free energy with respect to the independent variables gives equality of component fugacities among all phases for $(n_p - 1)\ n_c$ primary variables. The resultant equation that governs the flash calculations is as follows:

$$f_i^j - f_i^r = 0 \quad \left(i = 1, 2, \ldots, n_c;\ j = 1, 2, \ldots, n_p - 1 \right). \tag{6.46}$$

The phase composition constraint is defined as the following:

$$\sum_{i=1}^{n_c} x_{ij} - 1 = 0 \quad \left(j = 1, 2, \ldots, n_p \right), \tag{6.47}$$

and the equations for calculating the phase amounts for two hydrocarbon phases is defined as

$$\sum_{i=1}^{n_c} \frac{z_i(K_i - 1)}{1 + v(K_i - 1)} = 0, \tag{6.48}$$

$$K_i = \frac{y_i}{x_i}, \tag{6.49}$$

where z_i is the overall composition of component i, v is the ratio of moles of gas to total moles, and K_i values are defined as the ratio of the component mole fraction in a reference phase (gas phase, y_i) to that in another phase (oil phase, x_i).

Equations (6.45) and (6.46) are implicitly used in the solution of the fugacity equation, Equation (6.44). The volume constraint states that the total fluid volume must fill completely the pore volume in each of the cells, which is defined as follows:

$$\sum_{i=1}^{n_c} N_i \sum_{j=1}^{n_p} L_j \overline{v}_j - V_p = 0, \tag{6.50}$$

where L_j is a ratio of moles in phase j to the total number of moles in the mixture, $\overline{v}_j$ is the molar volume of phase j, and V_p is the pore volume of a gridblock.

Because of the IMPEC nature of the UTCOMP simulator, the gridblock pressure needs to be implicitly solved first. The pressure equation used in UTCOMP satisfies the condition that the pore volume should be filled completely by the total fluid volume:

$$V_t(P, \vec{N}) = V_P(P), \tag{6.51}$$

where the fluid is assumed a function of pressure and total number of moles of each component and the pore volume are related to pressure only. By differentiating both volumes with respect to time, using the chain rule to expand both terms against their independent variables, and then rearranging and substituting Equation (6.41) into the resultant equation, the final pressure equation becomes as follows:

$$\left(V_p^o c_f - \frac{\partial V_t}{\partial P}\right)\left(\frac{\partial P}{\partial t}\right) - V_b \sum_{i=1}^{n_c+1} \overline{V}_{ti} \vec{\nabla} \cdot \sum_{j=1}^{n_p} \vec{\vec{k}} \lambda_{rj} \xi_j x_{ij} \nabla P$$

$$= V_b \sum_{i=1}^{n_c+1} \overline{V}_{ti} \vec{\nabla} \cdot \sum_{j=1}^{n_p} \vec{\vec{k}} \lambda_{rj} \xi_j x_{ij} \left(\nabla P_{c2j} - \gamma_j \nabla D \right)$$
$$+ V_b \sum_{i=1}^{n_c+1} \overline{V}_{ti} \vec{\nabla} \cdot \sum_{j=1}^{n_p} \varphi \xi_j S_j \vec{\vec{K}} \nabla x_{ij} + \sum_{i=1}^{n_c+1} \overline{V}_{ti} q_i, \quad (6.52)$$

where V_p^o is the pore volume at the reference pressure P^o, c_f is the formation compressibility, P_{c2j} is the capillary pressure between phase 2, oil phase, and phase j, and $\overline{V}_{ti}$ is the partial molar volume of component i. Equation (6.50) is solved for pressure (P) at a given time t and the values of the rest of physical quantities are taken at the previous time level.

A simplified UTCOMP flowchart is shown in Fig. 6.2, where the simulation run starts at time t after the initialization step. In the initialization step, gridblock pressures are corrected for gridblock depths and phase behavior calculation is performed using the overall hydrocarbon mole fractions provided by the user.

More information about the UTCOMP simulator and the discretization of pressure and mass conservation equations can be found elsewhere (Chang, 1990).

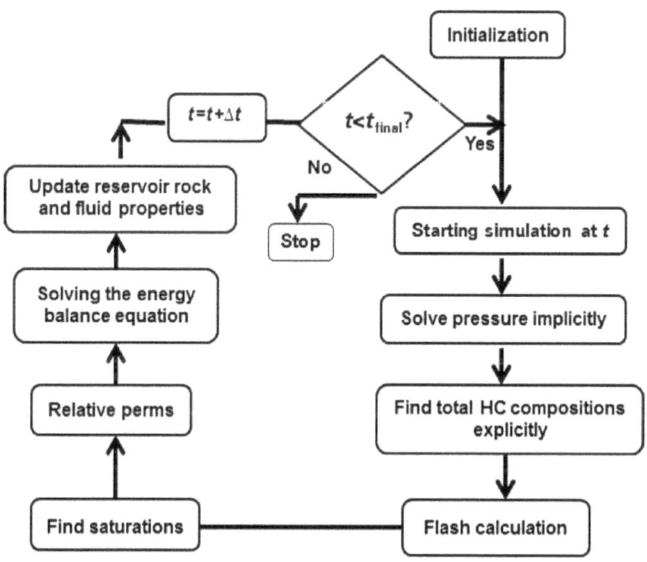

Figure 6.2 Simplified UTCOMP calculation flowchart (Korrani, 2014).

6.2.2 PHREEQC geochemical engine

PHREEQC (pH-REdox-Equilibrium in C programming language) is a free, open-source state-of-the-art geochemical package, developed by United States Geological Survey. It is a very flexible tool for modeling reactive-transport studies with rich databases. PHREEQC has capabilities for speciation and saturation index calculations, batch reaction and one-dimensional transport calculations with reversible and irreversible reactions including aqueous, mineral, gas, solid-solution, surface complexation, and ion-exchange equilibria, kinetically controlled reactions, multicomponent diffusion and dispersion, mixing of solutions, and inverse modeling (Parkhurst and Appelo, 2013).

The main equations used for aqueous speciation model in PHREEQC are mole balance equations, aqueous mass−action equations, activity coefficient model, and SI. Activity coefficient models are models describing the relation between the activity coefficients of species and ionic strength of the solution.

There are many generalized geochemical software/engines capable of modeling geochemical reactions, such as MINEQL+ (Schecher and McAvoy, 1992), Geochemist's Workbench (Bethke and Yeakel, 2009), PHREEQC (Parkhurst and Appelo, 2013), to list a few. These geochemical engines provide comprehensive reactive models. PHREEQC was chosen because it is a complete geochemical package that provides all the essential capabilities required for comprehensive reactive-transport modeling. Some of the features that distinguish this geochemistry engine from others include:

- PHREEQC can easily handle aqueous reactions in contact with a gas.
- Local equilibrium as well as kinetically controlled solid/fluid interactions are implemented in PHREEQC.
- PHREEQC has several activity coefficient models.

6.2.3 Implementation of geochemical species in UTCOMP and coupling with IPHREEQC

Korrani (2014) transcribed the transport of geochemical elements into the UTCOMP simulator. This means the mass conservation Equation (6.41) is now solved for both hydrocarbon and geochemical species at each time step. Rather than solving the mass conservation equation for all geochemical species, a common practice is to solve the equation just for the master geochemical species (elements) to accelerate the computational time

(Bhuyan, 1989). For instance, if an aqueous phase contains the following geochemical species: Na^+, Ca^{2+}, H^+, CO_3^{2-}, SO_4^{2-}, $CaCO_{3(aq)}$, HCO_3^-, $NaSO_4^-$, and $CaSO_{4(s)}$, then the master geochemical species (elements) are Na, Ca, S, C, H, and O. The latter geochemical elements are evaluated from the former geochemical species using the stoichiometric coefficients multiplied by the geochemical species concentration. In this case, the concentration of geochemical elements is represented by

$$C_{Ca(initial)} = C_{Ca^{2+}} + C_{CaCO_3\,(aq)} + C_{CaSO_{4(s)}}, \tag{6.53}$$

$$C_{Na(initial)} = C_{Na^+} + C_{NaSO_4^-}, \tag{6.54}$$

$$C_{S(6)(initial)} = C_{SO_4^{2-}} + C_{NaSO_4^-} + C_{CaSO_{4(s)}}, \tag{6.55}$$

$$C_{C(4)(initial)} = C_{CO_3^{2-}} + C_{CaCO_3\,(aq)} + C_{HCO_3^-}, \tag{6.56}$$

$$C_{H(initial)} = C_{H^+} + C_{HCO_3^-}. \tag{6.57}$$

At each time step, once the elements concentrations are calculated for gridblocks, these concentrations are fed to a geochemical engine (IPHREEQC) to find the equilibrium state for each gridblock. Although in reality, both transport and chemical reactions occur simultaneously, in this approach, the transport and the geochemistry are separated and solved in a sequential noniterative approach.

6.2.4 Batch reaction calculations

Geochemical species are active tracers that interact with each other though geochemical reactions. Therefore, after solving the mass balance equation for the new concentrations of geochemical elements, geochemical reactions between geochemical elements must be taken into consideration. Usually, each grid in the simulation model is considered as a batch cell where batch reactions occur; meaning each gridblock can be assumed as a beaker with a solution and set of reactants inside that are allowed to react (Zhu and Anderson, 2002).

An example on batch reactions is illustrated here by considering a beaker with a certain concentration of the previous geochemical elements (Na, Ca, S, C, H, and O). Also, we assume that the following geochemical reactions only exist between these elements in the aqueous phase:

$$Ca^{2+} + CO_3^{2-} \leftrightarrow CaCO_{3(aq)}, \tag{6.58}$$

$$H^+ + CO_3^{2-} \leftrightarrow HCO_3^-, \tag{6.59}$$

$$Na^+ + SO_4^{2-} \leftrightarrow NaSO_4^-. \tag{6.60}$$

We also consider a dissolution/precipitation reaction of anhydrite ($CaSO_{4(s)}$) as follows:

$$CaSO_{4(s)} \leftrightarrow Ca^{2+} + SO_4^{2-}. \tag{6.61}$$

In this case, in order to determine equilibrium state of the system, the concentrations of Na^+, H^+, Ca^{2+}, CO_3^{2-}, SO_4^{2-}, $CaCO_{3(aq)}$, HCO_3^-, $NaSO_4^-$, and $CaSO_{4(s)}$ should be known. Hence, nine equations are needed to define a unique solution for the equilibrium state of the batch cell.

Five equations are obtained from the mass material balance for each geochemical element as was previously discussed in Equations (6.51–6.55). Another three equations are obtained from the mass–action equations for the geochemical reactions in the aqueous phase as follows:

$$K_1 = \frac{\left[CaCO_{3(aq)}\right]}{\left[Ca^{2+}\right]\left[CO_3^{2-}\right]}, \tag{6.62}$$

$$K_2 = \frac{\left[HCO_3^-\right]}{[H^+]\left[CO_3^{2-}\right]}, \tag{6.63}$$

$$K_3 = \frac{\left[NaSO_4^-\right]}{[Na^+]\left[SO_4^{2-}\right]}. \tag{6.64}$$

The last equation is obtained from the solubility produce of the solid phase present in the system as the following:

$$K_{sp} \geq \left[Ca^{2+}\right]\left[SO_4^{2-}\right]. \tag{6.65}$$

In case the solid is present in the system, then the corresponding solubility product constraint is the equality; however, if the solid is absent, the constraint is inequality.

It should be noted that the above brackets represent the concentrations of geochemical species. Nevertheless, as was discussed previously in Equation (6.14), this is an approximation of assuming activity coefficients equal to unity, and hence activities of different geochemical species are equal to their concentrations. Moreover, it is also worth mentioning that equilibrium constants (K) and solubility products (K_{sp}) are temperature- and pressure dependent and are well documented in the literature for different reactions.

Equations (6.51–6.55) and (6.60–6.63) can be arranged as follows:

$$C_{Ca(initial)} - \left(C_{Ca^{2+}} + C_{CaCO_3\ (aq)} + C_{CaSO_{4(s)}}\right) = 0, \tag{6.66}$$

$$C_{Na(initial)} - \left(C_{Na^+} + C_{NaSO_4^-}\right) = 0, \tag{6.67}$$

$$C_{S(6)(initial)} - \left(C_{SO_4^{2-}} + C_{NaSO_4^-} + C_{CaSO_{4(s)}}\right) = 0, \tag{6.68}$$

$$C_{C(4)(initial)} - \left(C_{CO_3^{2-}} + C_{CaCO_{3(aq)}} + C_{HCO_3^-}\right) = 0, \tag{6.69}$$

$$C_{H(initial)} - \left(C_{H^+} + C_{HCO_3^-}\right) = 0, \tag{6.70}$$

$$K_1\left\{\gamma_{Ca^{2+}} C_{Ca^{2+}}\right\}\left\{\gamma_{CO_3^{2-}} C_{CO_3^{2-}}\right\} - \left\{\gamma_{CaCO_3\ (aq)} C_{CaCO_3\ (aq)}\right\} = 0, \tag{6.71}$$

$$K_2\left\{\gamma_{H^+} C_{H^+}\right\}\left\{\gamma_{CO_3^{2-}} C_{CO_3^{2-}}\right\} - \left\{\gamma_{HCO_3^-} C_{HCO_3^-}\right\} = 0, \tag{6.72}$$

$$K_3\left\{\gamma_{Na^+} C_{Na^+}\right\}\left\{\gamma_{SO_4^{2-}} C_{SO_4^{2-}}\right\} - \left\{\gamma_{NaSO_4^-} C_{NaSO_4^-}\right\} = 0, \tag{6.73}$$

$$K_{sp} \geq \left\{\gamma_{Ca^{2+}} C_{Ca^{2+}}\right\}\left\{\gamma_{SO_4^{2-}} C_{SO_4^{2-}}\right\}. \tag{6.74}$$

Equations (6.64–6.72) are highly coupled and nonlinear. Hence, iterative methods like the Newton–Raphson method can be used to find the solution for the set of these nonlinear equations. It is worth mentioning that the mass–action equations (6.69–6.72) are expressed in a general form in terms of activities of different geochemical species. In addition, activity coefficients for different species (γ) can be expressed through different activity coefficient models, which introduce more nonlinearity to the set of equations. It should be noted that charge balance should be considered as well; meaning that the sum of the positive and negative charges must be equal because real solutions must be charge balanced. Usually, one of the charge species, e.g., Cl^-, is adjusted to give the balance. For example, UTCHEM directly uses the charge balance equation of the aqueous species to take the place of the elemental mass balance equation for oxygen, while PHREEQC uses charge balance equation of the aqueous species as a criterion to check if the solution is neutralized, and otherwise the iteration continues.

Korrani (2014) coupled IPHREEQC to the UTCOMP simulator to apply the batch reaction calculations for each reservoir gridblock at each time step. IPHREEQC is a free and open-source module of the PHREEQC package designed to be used in scripting languages and integrated into C++, C, and Fortran programs (Charlton and Parkhurst, 2011). The "I" in IPHREEQC stands for "interface", hence the IPHREEQC provides an interface through which data transfer between UTCOMP and PHREEQC geochemical engine is achieved through a well-defined set of methods without writing and reading files. More information about the coupling of IPHREEQC with UTCOMP can be found elsewhere (Korrani, 2014).

Noteworthy, the approach presented above for batch calculations can be further expanded to include other batch reactions than those discussed (e.g., surfactant cation associated, surface complexation, kinetic, and exchange reactions).

6.2.5 Incorporating hydrocarbon phase effect on aqueous–rock geochemistry

Several studies have been reported in the literature on the importance of water-soluble hydrocarbon components on multiphase reactive-transport, especially in the field of CO_2 storage and sequestration (Nghiem et al., 2004; Liu and Maroto-Valer, 2010; Zhang et al., 2011; Xu et al., 2011). The effect of hydrocarbons on aqueous–rock geochemistry can be divided into two categories: soluble hydrocarbon components (e.g., CH_4 and CO_2) and acidic/basic components.

- *Soluble Hydrocarbon Components*

  In soluble hydrocarbon components, CO_2 dissolves in the aqueous phase and forms carbonic acid. The latter acid affects the pH of the system as well as the carbonate anion, which can interact with other cations present in the aqueous phase and form insoluble carbonates. Hence, soluble hydrocarbon components affect directly mineral dissolution and precipitation. Moreover, in an indirect way, the soluble hydrocarbon components can coat mineral surfaces and reduce the effective surface area of the mineral for precipitation and dissolution (Zhang and Villegas, 2012).

  From basic thermodynamics, soluble hydrocarbon components have identical fugacities in all existing phases (water, oil, and gas). This concept is usually used to model the solubility of hydrocarbon components as follows:

$$f^{w}_{CO_2} = f^{o}_{CO_2} = f^{g}_{CO_2}, \tag{6.75}$$

$$f^{w}_{CH_4} = f^{o}_{CH_4} = f^{g}_{CH_4}. \tag{6.76}$$

- *Acidic/Basic Components*

  On the other hand, acidic/basic components of crude oil can also exchange between the hydrocarbon and aqueous phases. The acidic components include mainly all carboxylic acids (RCOOH) which can be found in the resin and asphaltene fractions (Farooq et al., 2011). These components are extremely important in alkaline/surfactant/polymer floodings (Bhuyan, 1989). Basic components include heterocyclic aromatic compounds with one nitrogen atom and one or more aromatic rings (Farooq et al., 2011). Both acidic and basic components have been recently studied to interpret the underlying mechanism for incremental oil recovery by LSWI/EWI (Austad et al., 2010; RezaeiDoust et al., 2011; Farooq et al., 2011).

IPHREEQC/PHREEQC properly takes into account the effect of soluble hydrocarbon components on the aqueous–rock geochemistry. Nevertheless, Korrani (2014) modified IPHREEQC database to include the acidic/basic geochemical reactions along with other reactions in UTCOMP–IPHREEQC. There are two common approaches for considering the soluble hydrocarbon components into account: the sequential iterative approach (Mangold and Tsang, 1991) and the simultaneous solution method or fully coupled approach (Steefel and Lasaga, 1992). For the sequential iterative approach, flow equation and geochemistry equations are in an iteration loop and solved separately and sequentially until the convergence achieved. Nevertheless, for the simultaneous solution method, Newton's method is applied to solve the system of equations simultaneously. Korrani (2014) used the sequential iterative approach as it uses much less computer memory compared to the fully coupled approach. A simplified UTCOMP flowchart after coupling with IPHREEQC is depicted in Fig. 6.3, which includes both mass conservation equation for geochemical species and the hydrocarbon phase effect on the aqueous–rock geochemistry.

In the Fig. 6.3, TNK_i, $TNKTG_i$, and f_i are the total moles of the hydrocarbon component i, total moles of the geochemical element i, and fugacity of the component i, respectively. In this approach, the solubility

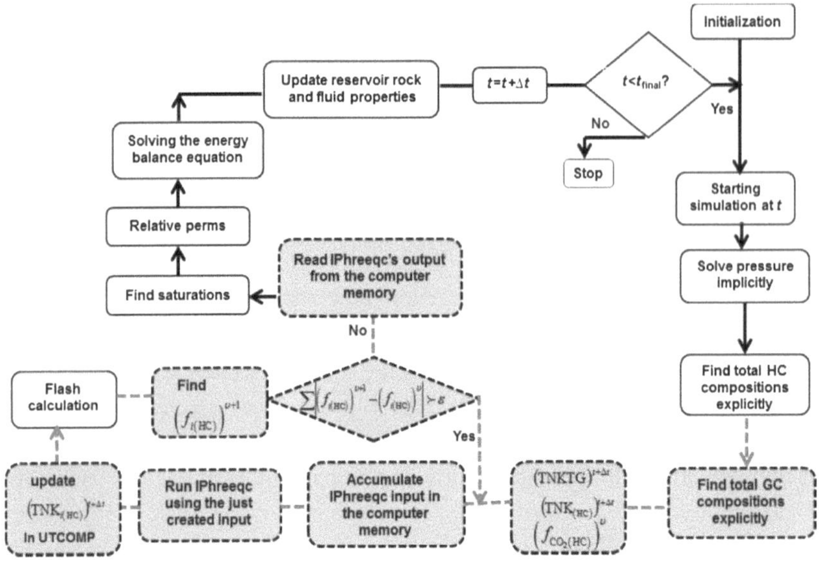

Figure 6.3 Simplified UTCOMP–IPHREEQC calculation flowchart with the hydrocarbon phase effect on the aqueous–rock geochemistry (Korrani, 2014).

of hydrocarbon phase components is taken into account by performing two sequential flash calculations (hydrocarbon–aqueous phase and oil–gas flash calculations).

Korrani (2014) proposed a simpler flowchart for the case of a single phase (only aqueous phase) or two phases (dead oil and aqueous phases) where the effect of the hydrocarbon phase on the aqueous–rock geochemistry is negligible (Fig. 6.4). In this case, the phase composition calculation of the hydrocarbon and aqueous phases is independent.

Further, Korrani (2014) parallelized the geochemistry module of the UTCOMP–IPHREEQC integrated simulator to reduce computational time, especially for large-scale reservoir simulations. More information on the computational algorithm of the parallel-processing version of UTCOMP–IPHREEQC can be found elsewhere (Korrani, 2014).

6.3 APPLICATION OF GEOCHEMISTRY IN LSWI/EWI

The effects of dissolution and precipitation processes on incremental oil by LSWI were modeled. Evje et al. (2009) proposed a 1D

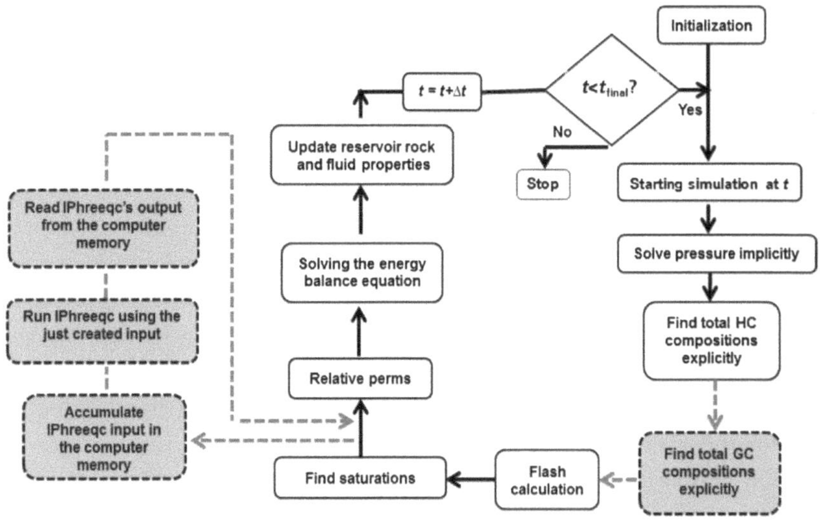

Figure 6.4 Simplified UTCOMP–IPHREEQC calculation flowchart without the hydrocarbon phase effect on the aqueous–rock geochemistry (Korrani, 2014).

mathematical model for the experimental results obtained by Madland (2009) by coupling dissolution/precipitation and transport processes relevant to chalk weakening effect in carbonate reservoirs. The model is composed of a number of convection-diffusion reaction equations represented in the water phase, coupled with ordinary differential equations represented the solid phase. The model was evaluated by comparison with experimental behavior in which a mixture of water and $MgCl_2$ was injected at constant rate in a core initially filled with pure water at high temperature 130°C. The main characteristics of the model were the precipitation of $MgCO_3$ and dissolution of $CaCO_3$ with injection of $MgCl_2$ to the chalk core.

Later, Evje and Hiorth (2009) proposed a novel mathematical model based on laboratory experiments, which considers two phases water–oil (pressure driven or capillary driven), aqueous chemistry and water–rock interactions, and dynamic wettability alteration due to water–rock interactions. The proposed model was a 1D model for spontaneous imbibition as wettability alteration was related to the dissolution of calcite. Dynamic wettability alteration was considered by defining relative permeability and capillary pressure curves for both water-wet and mixed-wet cases and interpolating in between depending on calcium dissolution with emphasis on concentrations of sulfate and magnesium ions. Moreover, the effect of temperature change was included in light of the observed experimental behavior. This model was considered different from other models, as it is one of the rare models including the two-phase behavior with emphasis on both the water–rock interactions along with modeling the oil recovery.

The latter model assumes that the rock is initially saturated with formation brine, which is at equilibrium with rock minerals. Once water with modified ionic composition is injected, the ions are transported to the rock by molecular diffusion forming concentration fronts moving with certain speed, and chemical interactions occur at and behind these fronts. These reactions are aqueous reactions for achieving chemical equilibrium and water–rock interactions, and the water–rock interactions are the one responsible for wettability alteration by dissolution of calcite. The model included various ion species concentrations and reactions occurring as result of water–rock interactions like

$$\begin{aligned} &CaCO_3(s) + H^+ \leftrightarrow Ca^{2+} + HCO_3^- \text{ (dissolution/precipitation)},\\ &CaSO_4(s) \leftrightarrow Ca^{2+} + SO_4^{2-} \text{ (dissolution/precipitation)},\\ &MgCO_3(s) + H^+ \leftrightarrow Mg^{2+} + HCO_3^- \text{ (dissolution/precipitation)}. \end{aligned} \tag{6.77}$$

In addition to the interactions in the aqueous phase

$$\begin{aligned} CO_2 + H_2O &\leftrightarrow HCO_3^- + H^+,\\ HCO_3^- &\leftrightarrow CO_3^{2-} + H^+, \end{aligned} \tag{6.78}$$

This model incorporated the chemical activity of various ions along with the definition of the activity coefficient. The flow functions of capillary pressure and relative permeability were expressed in terms of calcite dissolution ($H(\rho_c)$), respectively as the following:

$$\begin{aligned} P_c(s,\rho_c) &= H(\rho_c)P_c^{ow}(s) + [1 - H(\rho_c)]P_c^{ww}(s),\\ k(s,\rho_c) &= H(\rho_c)k^{ow}(s) + [1 - H(\rho_c)]k^{ww}(s). \end{aligned} \tag{6.79}$$

Moreover, Hiorth et al. (2010) investigated the effect of water chemistry on surface charge and rock dissolution in a pure calcium carbonate rock and constructed a chemical model that couples bulk aqueous and surface chemistry addressing mineral dissolution and precipitation. Spontaneous imbibition experiments on Stevns Klint outcrop chalk were used to compare the proposed model. The results showed clearly that the strong relation between oil recovery and temperature cannot be justified based on changes in the surface potential, as at high temperatures the surface potential levels off and thereof surface charge is not the main controlling factor for wettability alteration. The proposed justification behind oil recovery strong relation with temperature was calcite dissolution. Calcite is in equilibrium with seawater at low temperature; however, with increasing the temperature, the calcite and sulfate in seawater react resulting in anhydrite precipitation. Thus, loss in calcium from the aqueous phase requires calcium to be supplied from the rock for maintaining equilibrium with calcite. The source of Ca^{2+} supply is calcium dissolution. If calcium dissolution happens at the oil adsorption place, this leads to release of oil droplets.

Thus, unstable thermodynamic conditions at high temperatures were proposed as the reason behind wettability alteration in the form of calcite dissolution as opposed to changes in calcite surface charge, especially in justifying the increase in oil recovery when positively charge ions (Ca^{2+} and Mg^{2+}) injected to the cores. This study contradicts with the observations from NMR T_2 distribution curves by Yousef et al. (2011) which show that changing surface charge is more important than dissolution; otherwise, higher amplitudes can be obtained in NMR T_2 distribution curves indicating slower rate of relaxation.

In addition, Omekeh et al. (2012) developed a model that describes dissolution/precipitation processes in carbonate rock along with multicomponent ions exchange to simulate the additional oil recovery by LSWI/EWI. In this model, the release of divalent cations from the rock surface changes the relative permeability so that additional oil is achieved. A scaling function (F), which depends on desorption of the divalent ions, was defined for interpolating between predetermined high-salinity and low salinity sets of relative permeability as follows:

$$k\left(S,\beta_{\mathrm{Ca}},\beta_{\mathrm{Mg}}\right)=F\left(\beta_{\mathrm{Ca}},\beta_{\mathrm{Mg}}\right)k^{\mathrm{HS}}\ (\mathrm{S})+\left[1-F\left(\beta_{\mathrm{Ca}},\beta_{\mathrm{Mg}}\right)\right]k^{\mathrm{LS}}\ (\mathrm{S}), \tag{6.80}$$

as β_{Ca} and β_{Mg} are the absorbed calcium and magnesium cations, respectively. The suggested model was validated by history matching a two-phase coreflooding experiment. They concluded that calcite dissolution and ion exchange are responsible for the LSWI/EWI effect on oil recovery.

Few of the presented LSWI models capture the effect of different geochemical reactions on oil recovery. Dang et al. (2013) proposed a low salinity water model by coupling a comprehensive ion exchange model with geochemical processes to a multiphase multicomponent flow EOS compositional simulator. The interpolating factor they used to interpolate between relative permeability curves is defined as follows:

$$\omega = \frac{\mathrm{Ca} - \mathrm{X}_2 \times \mathrm{CEC}}{\mathrm{CEC}_{\mathrm{max}}}, \tag{6.81}$$

where $Ca{-}X_2$ is the equivalent fraction of Ca^{2+} on the clay exchanger, CEC is the cation exchange capacity of clays, and CEC_{max} is the maximum cation exchange capacity of clay. The proposed mode was validated using low salinity corefloods reported by Fjelde et al. (2012) for a North Sea reservoir and heterogeneous corefloods by Rivet (2009) for a Texas reservoir.

Also, Korrani et al. (2013) coupled IPHREEQC, the state-of-art geochemical engine (Parkhurst and Appelo, 2013), with UTCHEM to develop a robust, accurate, and flexible tool named as UTCHEM–IPHREEQC simulator. The latter simulator was used for modeling low salinity water injection and other geochemically related IOR/EOR processes. In a later work, Korrani et al. (2014) coupled IPHREEQC with UTCOMP as was discussed thoroughly above. The motive behind the latter coupling is to show the effect of water-soluble hydrocarbon components (e.g., CO_2, CH_4, and acidic/basic components of the crude oil) on buffering the aqueous pH and more generally, on the crude oil, brine, and rock reactions.

Korrani (2014) used the latter simulator for modeling the experimental work of Kozaki (2012) through using the total ionic strength as the interpolating parameter between two sets of relative permeability curves. The philosophy for this modeling was first proposed by Ligthelm et al. (2009) as follows:

$$\theta = \frac{\mathrm{TIS}_{\mathrm{max}} - \mathrm{TIS}(x,t)}{\mathrm{TIS}_{\mathrm{max}} - \mathrm{TIS}_{\mathrm{min}}}, \tag{6.82}$$

where θ is the interpolating parameter, TIS_{max} is the total ionic strength value above which there is no wettability alteration, $TIS(x,t)$ is the total ionic strength of each gridblock at certain simulation time, and TIS_{min} is the total ionic strength value at which maximum wettability alteration occurs.

Moreover, Korrani (2014) modeled one of the corefloods conducted by Chandrasekhar and Mohanty (2013) on carbonate cores through considering the following interpolating parameter:

$$\theta = \frac{\xi_{\max} - \xi(x, t)}{\xi_{\max} - \xi_{\min}}, \quad (6.83)$$

where $\xi(x, t)$ is the amount of calcite in each gridblock, $\xi_{\max}$ is the amount of calcite above which there is no wettability alteration, and $\xi_{\min}$ is the value at which enough amount of calcite is dissolved making the rock totally water-wet. It is worth mentioning that the geochemical model included aqueous, dissolution/precipitation, exchange, and surface complexation reactions.

Qiao et al. (2015) presented a mechanistic geochemical model for predicting incremental oil recovery by LSWI that includes surface complexation, aqueous reactions, and dissolution/precipitation of calcite and anhydrite. The geochemical reactions were coupled with the equations of multiphase flow and transport in an IMPEC in-house simulator, PennSim (PennSim Toolkit, Qiao et al., 2014). Relative permeability and residual oil situation are adjusted using the concentration of oil acids attached to the calcite mineral surface as the scaling factor. The scaling factor used as well as the corresponding chemical reactions are as follows:

$$\theta = \frac{c - c_{ww}}{c_{ow} - c_{ww}}, \quad (6.84)$$

$$>CaOH_2^+(-COO^-) \leftrightarrow >CaOH_2^+ + -COO^-, \quad (6.85)$$

where c, c_{ww}, and c_{wo} are the surface concentrations of $>CaOH_2^+(-COO^-)$ at the current state, the end-point water-wet state, and oil-wet state, respectively.

The proposed model successfully history matched four of the experimental corefloods in the literature (Strand et al., 2008; Fathi et al., 2010; Austad et al., 2012; Yousef et al., 2012). Moreover, they performed 2D five-spot patterns using tuned reaction networks where the incremental oil recovery ranged from 5% to 20% OOIP within reasonable values of pore volumes injected (2 PVI).

The importance of geochemical reactions and activity coefficient models was highlighted by Al-Shalabi et al. (2015). They presented a comparison between two geochemical simulators, UTCHEM (UTCHEM Technical Documentation, 2000) and PHREEQC

(Parkhurst and Appelo, 2013), through modeling fluid and solid species concentrations of a recently published LSWI coreflood (Yousef et al., 2011). Examples of fluid species comparison (sulfate) and solid species comparison (anhydrite) between both simulators are depicted in Figs. 6.5 and 6.6, respectively.

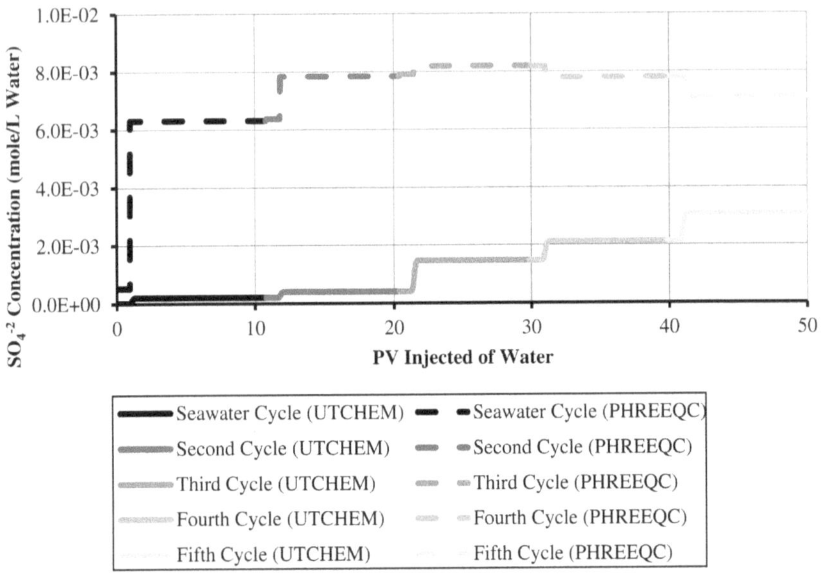

Figure 6.5 Sulfate ion concentration using UTCHEM and PHREEQC-fluid species (Al-Shalabi et al., 2015).

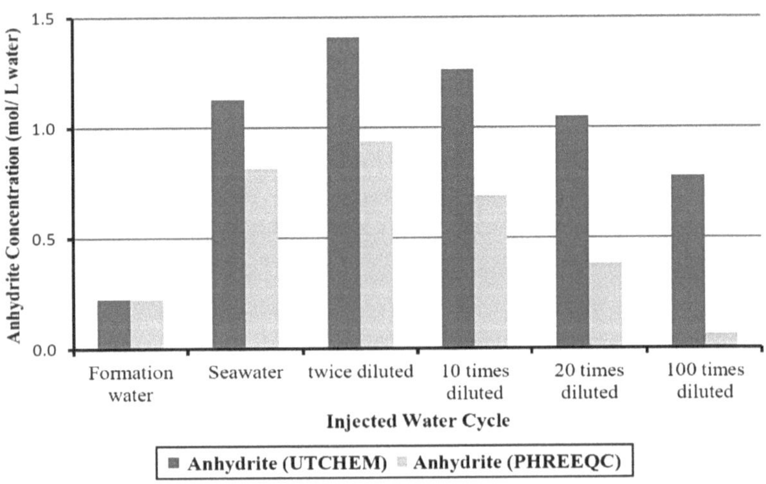

Figure 6.6 Anhydrite concentration at different injection cycles (Al-Shalabi et al., 2015).

In the latter work, the authors highlighted the necessity of considering activity coefficient models and water activity for accurate geochemical predictions of fluid- and solid-species concentrations. Based on the previous findings, Luo et al. (2016) improved the geochemical-reactive engine within UTCHEM by implementing different activity coefficient models for different reactive species, cation exchange reactions, and numerical convergence. The updated geochemical simulator was validated against both PHREEQC and UTCHEM–IPHREEQC showing better computational efficiency with the same degree of accuracy.

Al-Shalabi et al. (2016) utilized the latter updated version of UTCHEM through proposing a mechanistic LSWI model, which addresses the effect of different geochemical reactions resulting from LSWI on oil recovery. The proposed model was validated and used to history match recently published corefloods (Yousef et al., 2011 and 2012; Chandrasekhar and Mohanty, 2013). The geochemical model in UTCHEM was modified to calculate the effective molar Gibbs free energy of the brine. The molar Gibbs free energy of a solution is defined as follows (Sandler, 2006):

$$\underline{G} = \sum_{i=1}^{N_{\mathrm{aq}}} \overline{x}_i \mu_i, \tag{6.86}$$

where $\overline{x}_i$ is the mole fraction of each aqueous specie and μ_i is the chemical potential of each specie. The mole fraction and chemical potential of each species are expressed as

$$\overline{x}_i = \frac{x_i}{x_{\mathrm{tot}}}, \tag{6.87}$$

$$\mu_i = G_i^{\mathrm{o}} + RT \ln a_i, \tag{6.88}$$

where x_i is the number of moles for the species i, x_{tot} is the total number of moles in the solution, G_i^{o} is the standard Gibbs free energy, R is the universal gas constant, T is the temperature, and a_i is the activity of each specie. An example of effective molar Gibbs free energy calculated for the corefloods of Yousef et al. (2011) is depicted in Fig. 6.7.

In the proposed model, the relative permeability curves including the residual oil saturation are functions of the effective molar Gibbs free

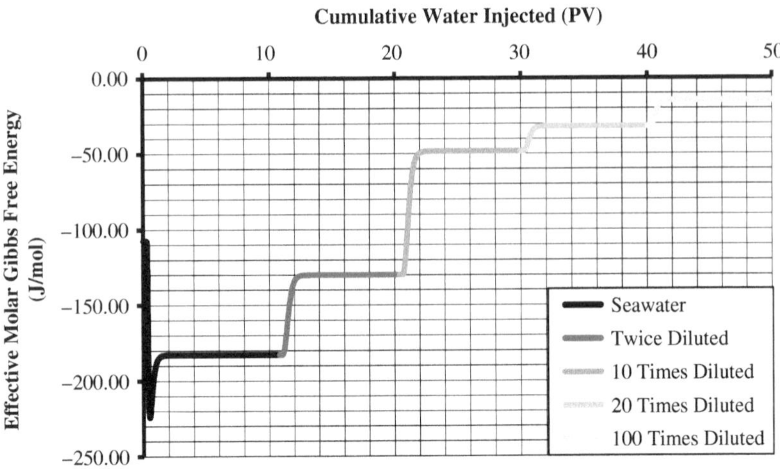

Figure 6.7 Effective molar Gibbs free energy calculated for corefloods of Yousef et al. (2011) using the mechanistic LSWI model (Al-Shalabi et al., 2016).

energy. The scaling factor used in this case for adjusting residual oil saturation as a function of effective molar Gibbs free energy is defined as

$$\omega S = \frac{\left(\underline{G} - \underline{G}^{HS}\right)}{\left(\underline{G}^{LS} - \underline{G}^{HS}\right)}, \tag{6.89}$$

where $\underline{G}$ is the effective molar Gibbs free energy (J/mol) at in-situ conditions of injected-connate mixed solution, $\underline{G}^{HS}$ is the effective molar Gibbs free energy (J/mol) at in-situ conditions of seawater-connate mixed solution, and $\underline{G}^{LS}$ is the effective molar Gibbs free energy (J/mol) at in-situ conditions of low salinity water—connate mixed solution when S_{or} becomes constant. An example of history matching effluent sulfate ion concentration for Chandrasekhar and Mohanty (2013) using the mechanistic LSWI model is shown in Fig. 6.8. The mechanistic model captures the effects on oil recovery of both wettability alteration and/or dissolution/fine migration through changes in the molar Gibbs free energy.

Mechanistic modeling of engineered water injection (EWI) effect on oil recovery from carbonate cores was performed by Adegbite et al. (2017a) through history matching the experimental work of Chandrasekhar and Mohanty (2013). CMG-GEM was used to history match and to model the effect of seawater with spiked sulfate

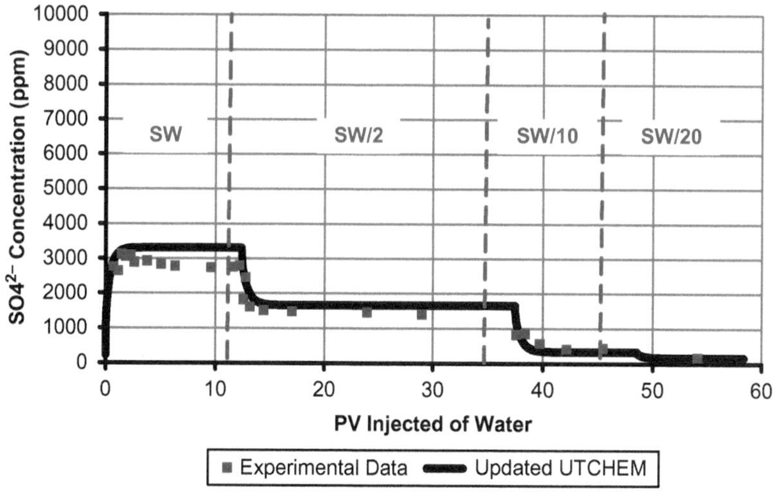

Figure 6.8 History matched sulfate concentration for coreflood of Chandrasekhar and Mohanty (2013) using the mechanistic LSWI model (Al-Shalabi et al., 2016).

concertation on oil recovery from a carbonate core. The core model was created using published rock-fluid and recovery data of a Middle-Eastern carbonate reservoir. Different geochemical reactions were captured in the simulation runs including aqueous, dissolution/precipitations, and ion-exchange reactions. An EWI model was proposed based on multiion exchange reactions for capturing incremental oil recovery by EWI in carbonates.

The proposed model is based on Zhang et al. (2006) mechanism of multiion exchange. For carbonate rocks an anion-exchange occurs on the positively charged carbonate surface, where the sulfate anion in water exchanges with the negatively charged carboxylic acid group in oil. This exchange results in releasing oil ganglia and altering the rock wettability towards a more water-wet state. With increasing the reservoir temperature the sulfate adsorption on rock surface increases. The chemical reaction for this proposed model is given below:

$$SO_4^{2-} + 2CH_3COO - X \leftrightarrow 2CH_3COO^- + SO_4 - X_2, \tag{6.90}$$

where X represents the carbonate rock and CH_3COO^- represents the carboxylic acid group of the oil. The latter equation was used in GEM for modeling the incremental oil recovery by EWI where the interpolation between relative permeability and capillary pressure curves for both formation water and engineered water (seawater with spiked sulfate) were

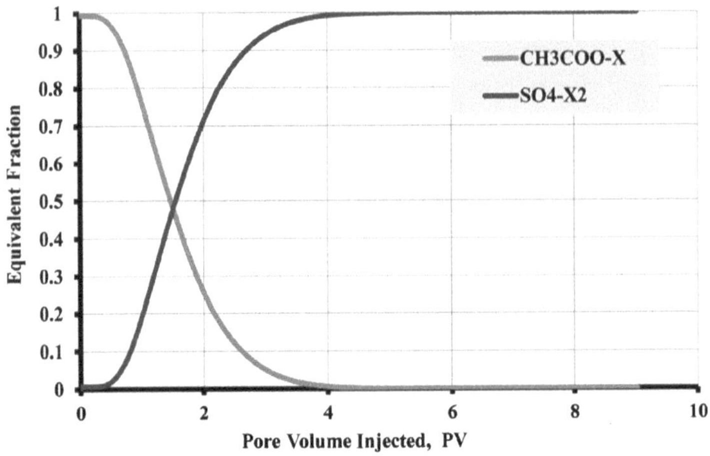

Figure 6.9 Equivalent fraction of SO_4-X_2 and CH_3COO-X at gridblock (1, 1, 1) (Adegbite et al., 2017a).

made using the equivalent fraction of sulfate on the carbonate surface ($(\zeta(SO_4{-}X_2))$ as the scaling factor.

Fig. 6.9 shows that the injection of seawater with spiked sulfates results in increasing SO_4^{2-} concentration and decreasing CH_3COO- concentration on the exchanger of the carbonate rock. Hence, in terms of equivalent fractions $\zeta(SO_4-X_2)$ increases while $\zeta(CH_3COO-X)$ decreases with time.

The released carboxylate ion (CH_3COO-) by the exchange reaction starts to form complexes with other free cations in the solution, e.g., ($CaCH_3COO^+$). The latter results in decreasing Ca^{2+} concentration in the solution, hence calcite dissolution occurs and pH increases. The dissolution and precipitation of minerals changes the void volume of the porous medium, and hence that was captured in the proposed model. Calcite dissolution results in enhancing the porosity of the modeled core (Fig. 6.10). Moreover, the Kozeny-Carman equation (Eq. (13)) was used to capture the change in permeability as the mineral dissolution/precipitation occurs (Fig. 6.11).

$$\frac{k}{k^0} = \left(\frac{\varphi}{\varphi^0}\right)^3 \cdot \left(\frac{1-\varphi^0}{1-\varphi}\right)^2, \tag{6.91}$$

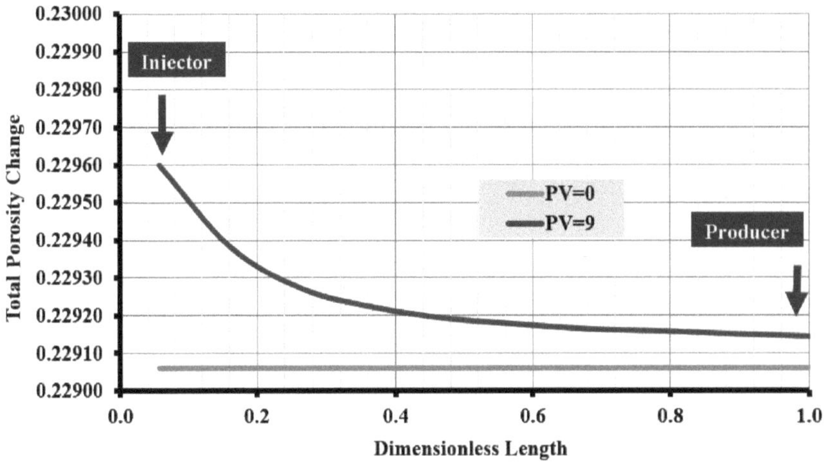

Figure 6.10 Total change in porosity profile (Adegbite et al., 2017b).

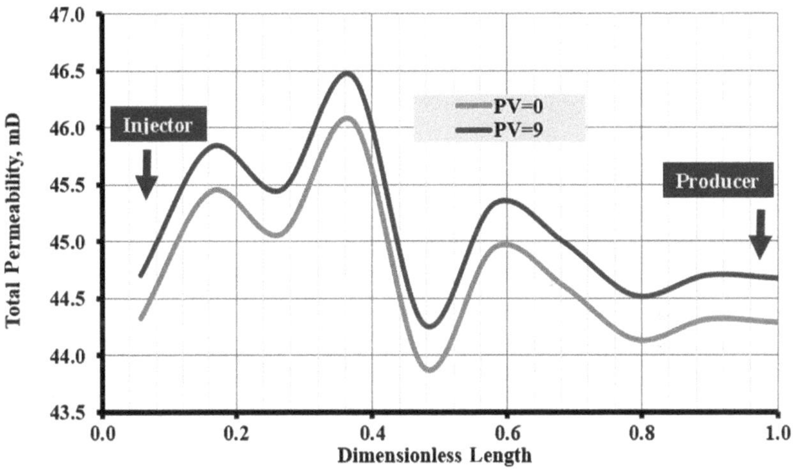

Figure 6.11 Total change in permeability profile (Adegbite et al., 2017b).

where k^0 and φ^0 are the initial permeability and porosity, respectively. Fig. 6.11 shows the enhancement in the absolute permeability after about 9 pore volumes as mineral dissolution occurs. It should also be noted that the total change in porosity and permeability profiles is more pronounced around the injector well compared to the producer well.

It is worth mentioning that oil recovery, pressure drop, and effluent ion concentrations data including pH were successfully history matched using the proposed model.

Other applications of this geochemical and compositional approach are discussed next in Chapter 7, which includes other applications of LSWI/EWI, especially the combined low salinity water and carbon dioxide injections.

REFERENCES

Adegbite, J.O., Al-Shalabi, E.W., Ghosh, B., 2017a. Modeling the effect of engineered water injection on oil recovery from carbonate cores. Paper SPE 184505, SPE International Conference on Oilfield Chemistry, Montgomery, Texas, USA.

Adegbite, J.O., Al-Shalabi, E.W., Ghosh, B., 2017b. Private communication.

Acs, G., Deleschall, S., Farkas, E., 1985. General purpose compositional model. SPE J. 25 (4), 543–553.

Al-Shalabi, E.W., Sepehrnoori, K., Delshad, M., 2014. Mechanisms behind low salinity water injection in carbonate reservoirs. Fuel J. 121, 11–19.

Al-Shalabi, E.W., Sepehrnoori, K., Pope, G., 2015. Geochemical interpretation of low salinity water injection in carbonate oil reservoirs. SPE J. 20 (6), 1212–1226.

Al-Shalabi, E.W., Sepehrnoori, K., Pope, G., 2016. Mechanistic modeling of oil caused by low-salinity–water injection in oil reservoirs. SPE J. 21 (3), 730–743.

Austad, T., RezaeiDoust, A., Puntervold, T., 2010. Chemical mechanism of low salinity water flooding in sandstone reservoirs. In: Paper SPE 129767, SPE Improved Oil Recovery Symposium, Tulsa, OK.

Austad, T., Shariatpanahi, S.F., Strand, S., Black, C.J.J., Webb, K.J., 2012. Conditions for a low-salinity enhanced oil recovery (EOR) effect in carbonate oil reservoirs. Energy Fuels. 26 (1), 569–575.

Bethke, C., Yeakel, S., 2009. Geochemist's Workbench: Release 8.0 Reference Manual. Champaign, IL.

Bhuyan, D., 1989. Development of an alkaline/surfactant/polymer compositional reservoir simulator. PhD Dissertation, The University of Texas at Austin, Austin, TX.

Chandrasekhar, S., Mohanty, K.K., 2013. Wettability alteration with brine composition in high temperature carbonate reservoirs. In: Paper SPE 166280, SPE Annual Technical Conference and Exhibition, New Orleans, LA.

Chang, Y., 1990. Development of a three-dimensional, equation-of-state compositional reservoir simulator for miscible gas flooding. PhD Dissertation, The University of Texas at Austin, Austin, TX.

Charlton, S.R., Parkhurst, D.L., 2011. Modules based on the geochemical model PHREEQC for use in scripting and programming languages. Comput. Geosci. 37 (10), 1653–1663.

Computer Modeling Group (CMG), 2016. User technical manual.

Dang, C.T.Q., Nghiem, L.X., Chen, Z., Nguyen, Q.P., 2013. Modeling low salinity waterflooding: ion exchange, geochemistry and wettability alteration. In: Paper SPE 166447, SPE Annual Technical Conference and Exhibition, New Orleans, LA.

Davis, J.A., James, R.O., Leckie, J.O., 1978. Surface ionization and complexation at the oxide/water interface. Computation of electrical double layer properties in simple electrolyte. Colloids Interfacial Sci. 63, 480–499.

Delshad, M., Pope, G.A., Sepehrnoori, K., 1996. A compositional simulator for modeling surfactant enhanced aquifer remediation. J. Contam. Hydrol. 23, 303–327.

Dzombak, D.A., Morel, F.M.M., 1990. Surface Complexation Modeling: Hydrous Ferric Oxide. John Wiley & Sons, New York.
Evje, S., Hiorth, A., 2009. A mathematical model for dynamic wettability alteration controlled by water–rock chemistry. Netw. Heterogen. Media. 5 (2), 217–256.
Evje, S., Hiorth, A., Madland, M.V., Korsnes, R.I., 2009. A mathematical model relevant for weakening of chalk reservoirs due to chemical reactions. Netw. Heterogen. Media. 4 (4), 755–788.
Farajzadeh, R., Matsuura, T., Van Batenburg, D., Dijk, H., 2012. Detailed modeling of the alkali/surfactant/polymer (ASP) process by coupling a multipurpose reservoir simulator to the chemistry package PHREEQC. SPE Reserv. Eval. Eng. 15 (4), 423–435.
Farooq, U., Asif, N., Tweheyo, M.T., Sjoblom, J., Oye, G., 2011. Effect of low salinity aqueous solutions and pH on the desorption of crude oil fractions from silica surfaces. Energy Fuels. 25 (5), 2058–2064.
Fathi, S.J., Austad, T., Strand, S., 2010. Smart water as a wettability modifier in chalk: the effect of salinity and ionic composition. Energy Fuels. 24 (4), 2514–2519.
Fjelde, I., Asen, S.M., Omekeh, A., 2012. Low salinity water flooding experiments and interpretation by simulations. In: Paper SPE 154142, SPE Improved Oil Recovery Symposium, Tulsa, OK.
Gaines, G.L., Thomas, H.C., 1953. Adsorption studies on clay minerals. A formulation of the thermodynamics of exchange adsorption. J. Chem. Phys. 21, 714–718.
Garrels, R.M., Christ, C.L., 1965. Solutions, Minerals, and Equilibria. Harper & Row, New York.
Green, D.W., Willhite, G.P., 1998. Enhanced Oil Recovery. SPE, Richardson, TX.
Grenthe, I., Plyasunov, A.V., Spahiu, K., 1997. Estimations of medium effects on thermodynamic data. Modelling in Aquatic Chemistry., OECD Publications, Chapter IX.
Havre, T.E., Sjoblom, J., Vindstad, J.E., 2003. Oil/water-partitioning and interfacial behavior of naphthenic acids. J. Dispers. Sci. Technol. 24 (6), 789–801.
Hiorth, A., Cathles, L.M., Madland, M.V., 2010. Impact of pore water chemistry on carbonate surface charge and oil wettability. Transp. Porous Media. 85 (1), 1–21.
Hirasaki, G.J., 1982. Interpretation of the change in optimal salinity with overall surfactant concentration. SPE J. 22 (6), 971–982.
Jhaveri, B.S., Youngren, G.K., 1988. Three parameter modification of the Peng Robinson equation of state to improve volumetric predictions. SPE Reserv. Eng. 3 (3), 1033–1040.
Korrani, A.K.N., 2014. Mechanistic modeling of low salinity water injection. PhD Dissertation, The University of Texas at Austin, TX.
Korrani, A.K.N., Sepehrnoori, K., Delshad, M., 2013. A novel mechanistic approach for modeling low salinity water injection. In: Paper SPE 166523, SPE Annual Technical Conference and Exhibition, LA.
Korrani, A.K.N., Jerauld, G.R., Sepehrnoori, K., 2014. Coupled geochemical-based modeling of low salinity waterflooding. In: Paper SPE 169115, SPE Improved Oil Recovery Symposium, Tulsa, OK.
Kozaki, C., 2012. Efficiency of low salinity polymer flooding in sandstone cores. Master Thesis, The University of Texas at Austin, Austin, TX.
Lager, A., Webb, K.J., Black, C.J.J., Singleton, M., Sorbie, K.S., 2008. Low salinity oil recovery: an experimental investigation. Petrophysics. 49 (1), 28–35.
Lager, A., Webb, K., Seccombe, J., 2011. Low salinity waterflood, Endicott, Alaska: geochemical study & field evidence of multicomponent ion exchange. In: 16th European Symposium on Improved Oil Recovery, Cambridge, England.
Lake, L.W., 1989. Enhanced Oil Recovery. Prentice Hall, Englewood Cliffs, NJ.
Ligthelm, D.J., Gronsveld, J., Hofman, J., Brussee, N., Marcelis, F., van der Linde, H., 2009. Novel waterflooding strategy by manipulation of injection brine composition.

In: Paper SPE 119835, SPE EUROPEC/EAGE Conference and Exhibition, Amsterdam, The Netherlands.
Liu, Q., Maroto-Valer, M.M., 2010. Investigation of the pH effect of a typical host rock and buffer solution on CO_2 sequestration in synthetic brines. Fuel Process. Technol. 91 (10), 1321–1329.
Lohrenz, J., Bray, B.G., Clark, C.R., 1964. Calculating viscosities of reservoir fluids form their compositions. J. Petrol. Technol. 16 (10), 1171–1176.
Luo, H., Al-Shalabi, E.W., Delshad, M., Sepehrnoori, K., 2016. A robust geochemical simulator to model improved oil recovery methods. SPE J. 21 (1), 55–73.
Madland, M.V., 2009. Rock–fluid interactions in chalk exposed to seawater, $MgCl_2$, and NaCl brines with equal ionic strength. In: 15th European Symposium on Improved Oil Recovery, Paris, France.
Malmberg, C.G., Maryott, A.A., 1956. Dielectric constant of water from 0°C to 1000°C. J. Res. Nat. Bureau Standards. 56, 1–8.
Mangold, D.C., Tsang, C.F., 1991. A summary of subsurface hydrological and hydrochemical models. Rev. Geophys. 29 (1), 51–79.
Manov, G.G., Bates, R.G., Hamer, W.J., Acree, S.F., 1943. Values of the constants in the Debye–Hückel equation for activity coefficients. J. Am. Chem. Soc. 65 (9), 1765–1767.
Martin, F.D., Oxley, J.C., Lim, H., 1985. Enhanced recovery of a "J" sand crude oil with a combination of surfactant and alkaline chemicals. In: Paper SPE 14295, SPE Annual Technical Conference and Exhibition, Las Vegas, NV.
Nelson, R.C., Pope, G.A., 1978. Phase relationships in chemical flooding. SPE J. 18 (5), 325–338.
Nelson, R.C., Lawson, J.B., Thigpen, D.R., Stegemeier, G.L., 1984. Cosurfactant-enhanced alkaline flooding. In: Paper SPE 12672, SPE Enhanced Oil Recovery Symposium, Tulsa, OK.
Nghiem, L., Sammon, P., Grabenstetter, J., Ohkuma, H., 2004. Modeling CO_2 storage in aquifers with a fully-coupled geochemical EOS compositional simulator. In: Paper SPE 89474, SPE Improved Oil Recovery, Tulsa, OK.
Omekeh, A., Friis, H.A., Fjelde, I., Evje, S., 2012. Modeling of ion-exchange and solubility in low salinity water flooding. In: Paper SPE 154144, SPE Improved Oil Recovery Symposium, Tulsa, OK.
Parkhurst, D.L., Appelo, C.A.J., 2013. Description of input and examples for PHREEQC version 3—a computer program for speciation, batch-reaction, one-dimensional transport, and inverse geochemical calculations. Modeling Techniques (Chapter 43 of Section A Groundwater, Book 6).
Peng, D.Y., Robinson, D.B., 1976. A new two-constant equation of state. Ind. Eng. Chem. Fundam. 15 (1), 59–64.
Pitzer, K.S., 1991. Ion interaction approach: theory and data correlation. Activity Coefficients in Electrolyte Solutions. CRC Press, Boca Raton, FL.
Qiao, C., Li, L., Johns, R.T., Xu, J., 2014. Compositional modeling of reaction-induced injectivity alteration during CO_2 flooding in carbonate reservoirs. In: Paper SPE 170930, SPE Annual Technical Conference and Exhibition, Amsterdam, The Netherlands.
Qiao C., Johns, R., Li, L., Xu, J., 2015. Modeling low salinity waterflooding in mineralogically different carbonates. In: Paper SPE 175018, SPE Annual Technical Conference and Exhibition, Houston, TX.
Reid, R.C., Prausnitz, J.M., Sherwood, T.K., 1987. The Properties of Gases and Liquids. fourth ed. McGraw-Hill, New York.
RezaeiDoust, A., Puntervold, T., Austad, T., 2011. Chemical verification of the EOR mechanism by using low saline/smart water in sandstone. Energy Fuels. 25 (5), 2151–2162.

Rivet, S., 2009. Coreflooding oil displacements with low salinity brine. Master of Science Thesis, University of Texas at Austin, Austin, TX.
Sandler, S.I., 2006. Chemical, Biochemical, and Engineering Thermodynamics. fourth ed John Wiley & Sons, New York.
Schecher, W.D., McAvoy, D.C., 1992. MINEQL+ : a software environment for chemical equilibrium modeling. Comput., Environ. Urban Syst. 16 (1), 65–76.
Shakiba, M., 2014. Modeling and simulation of fluid flow in naturally and hydraulically fractured reservoirs using embedded discrete fracture model (EDFM). MS Thesis, The University of Texas at Austin, Austin, TX.
Sheng, J.J., 2013. A comprehensive review of alkaline–surfactant–polymer (ASP) flooding. In: Paper SPE 165358, SPE Western Regional & AAPG Pacific Section Meeting 2013 Joint Technical Conference, Monterey, CA.
Steefel, C.I., Lasaga, A.C., 1992. Putting transport into water–rock interaction models. J. Geol. 20 (8), 680–684.
Strand, S., Austad, T., Puntervold, T., 2008. Smart water for oil recovery from fractured limestone: a preliminary study. Energy Fuels. 22 (5), 3126–3133.
Truesdell, A. H., and Jones, B., 1974. WATEQ, a computer program for calculating chemical equilibria of natural waters. U.S. Geological Survey, version 2, pp. 233–274.
Turek, E.A., Metcalfs, R.S., Yarborough, L., Robinson, R.L., 1984. Phase equilibria in CO_2-multicomponent hydrocarbon systems: experimental data and an improved prediction technique. SPE J. 24 (3), 308–324.
UTCHEM—9.0 Technical Documentation, 2000. The University of Texas at Austin, Volume II, Austin, TX.
UTCOMP—3.8 Technical Documentation, 2003. The University of Texas at Austin, Austin, TX.
Van't Hoff, J.H., 1884. Etudes de Dynamique Chrimique. Muller, Amsterdam, pp. 114–118.
Xu, T., Zheng, L., Tian, H., 2011. Reactive transport modeling for CO_2 geological sequestration. J. Petrol. Sci. Eng. 78 (3), 765–777.
Yousef, A.A., Al-Saleh, S., Al-Kaabi, A., Al-Jawfi, M., 2011. Laboratory investigation of the impact of injection-water salinity and ionic content on oil recovery from carbonate reservoirs. SPE Reserv. Eval. Eng. 14 (5), 578–593.
Yousef, A.A., Al Saleh, S., Al Jawfi, M., 2012. Improved/enhanced oil recovery from carbonate reservoirs by tuning injection water salinity and ionic content. In: Paper SPE 154076, SPE Improved Oil Recovery Symposium, Tulsa, OK.
Zhang, P., Tweheyo, M.T., Austad, T., 2006. Wettability alteration and improved oil recovery in chalk: the effect of calcium in the presence of sulfate. Energy Fuels. 20 (5), 2056–2062.
Zhang, W., Li, Y., Omambia, A.N., 2011. Reactive transport modeling of effects of convective mixing on long-term CO_2 geological storage in deep saline formations. Int. J. Greenhouse Gas Control. 5 (2), 241–256.
Zhang, G., Villegas, E.I., 2012. Geochemical reactive transport modeling in oil & gas industry—business drivers, challenges and solutions. In: TOUGH Symposium, Berkeley, CA.
Zhu, C., Anderson, G., 2002. Environmental Applications of Geochemical Modeling. Cambridge University Press, Cambridge, UK.

CHAPTER SEVEN

Synergy of LSWI/EWI and Other EOR Processes

Contents

Low salinity/engineered water injection is an emerging technology with other useful applications, such as conformance control, the combined effects of LSWI/EWI and polymer flooding, LSWI/EWI and surfactant flooding, and LSWI/EWI and CO_2 flooding. Ayirala and Yousef (2015) summarized the importance of injection-water chemistry in different EOR/IOR processes (Fig. 7.1). The different applications of LSWI/EWI are discussed in this chapter for both sandstone and carbonate rocks.

7.1 CONFORMANCE CONTROL APPLICATION

Low salinity/engineered water injection techniques can be used in conformance control processes by decreasing the salinity gradient with sufficient amounts of Ca^{2+} in the injected water. The injected Ca^{2+} mobilizes clay minerals, plugs the porous media, and reduces the absolute permeability in the thief zones. Hence, the injected fluid is diverted into low permeability zones, resulting in improving oil recovery in these zones. Nevertheless, there is a lack of experimental evidence to support the water blockage by LSWI/EWI (Dang et al., 2014).

Low Salinity and Engineered Water Injection for Sandstone and Carbonate Reservoirs.
DOI: http://dx.doi.org/10.1016/B978-0-12-813604-1.00007-9

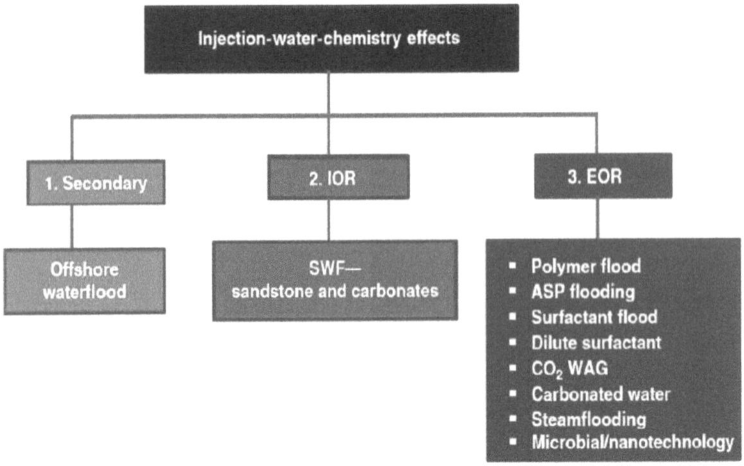

Figure 7.1 Summary of injection-water chemistry in different EOR/IOR processes (Ayirala and Yousef, 2015).

7.2 HEAVY OIL APPLICATION

Alzayer and Sohrabi (2013) performed several numerical simulations to investigate the potential of low salinity and low salinity augmented by polymer on heavy oil recovery using a sector model. The results showed an incremental oil recovery of approximately 5% OOIP using low salinity flooding alone; however, similar incremental recovery was obtained using polymer flooding alone with much better efficiency. The combination of low salinity and polymer floodings resulted in additional oil recovery of 7.5%–10% OOIP. The authors stressed on the need for laboratory work to validate the previous findings using heavy oils.

Jose et al. (2015) presented experimental and numerical works for evaluating the effect of LSWI on heavy-oil recovery from carbonates. The LSWI was conducted in both secondary and tertiary modes of injection. Results highlighted the potential of LSWI in recovering heavy-oil from carbonates where an oil recovery of 70% OOIP was obtained in the secondary mode. Moreover, the authors performed sensitivity analysis on different parameters that might affect the performance of LSWI in heavy-oil carbonate reservoirs. The authors found that seawater injection period before commencing LSWI is important as well as the water injection rate.

7.3 LSWI/EWI AND POLYMER FLOODING APPLICATION

Ayirala et al. (2010) addressed the advantages of using low salinity water as makeup water for polymer flooding. One of these advantages is the lower capital and operational costs compared to seawater, which is related to polymer facilities as the use of low salinity water reduces the need for more chemicals to meet a certain viscosity requirement. Moreover, a higher oil recovery can be obtained by improving the microscopic displacement efficiency using low salinity waterflooding and macroscopic sweep efficiency using polymer flooding. Kozaki (2012) conducted several corefloods in the tertiary mode on Berea sandstone cores using high-salinity/low salinity polymer solutions. He reported a decrease in residual oil saturation due to low salinity polymer solution by 5%–10% as opposed to the high-salinity polymer solution. Moreover, he highlighted the role of low salinity polymer solutions in achieving faster oil recovery by improving sweep efficiency. The combination of low salinity waterflooding and polymer is very attractive as one third or less of polymer is required for polymer floods, added to the 5-times reduction in chemical cost per barrel of oil recovered (Mohammadi and Jerauld, 2012).

Also, Vermolen et al. (2014) reported that by using low salinity polymer flooding, the project economics could improve as the polymer concentration for achieving a certain viscosity decreases by a factor two to four. Moreover, they highlighted additional benefits for using low salinity polymer injection including lower sensitivity to mechanical shear, higher stability and usage for high temperature/high-salinity formations, less potential production chemistry issues (scaling, souring, and oil/water separation), and possibly an additional reduction in residual oil saturation by increasing the visco-elasticity of the polymer solution. In addition, they stressed on the risks involved, including cation exchange, clay swelling, mixing of low salinity and high-salinity brines, and polymer adsorption. The additional polymer retardation factor caused by using a low salinity water leads to delay in oil recovery and might hurt the project economics. In addition, Han and Lee (2014) performed a sensitivity analysis for low salinity polymer flooding by investigating the effects of low salinity water slug size, water salinity, and polymer viscosity. The results showed that both LSWI slug size and polymer viscosity have a pronounced effect on low salinity polymer flooding. Nevertheless, a negligible effect was noticed for water salinity on the process.

7.4 LSWI/EWI AND SURFACTANT FLOODING APPLICATION

Spildo et al. (2012) investigated the effect of combining low salinity water injection and reduced capillarity by surfactants in terms of EOR additional oil recovery and cost efficiency. If capillary pressure is high, then retrapping of the LSWI mobilized oil might occur. Hence, surfactants are recommended to lower the capillary forces and avoiding retrapping of the mobilized oil. In the latter study, coreflooding experiments were conducted on Berea sandstone core plugs. The results concluded that surfactant injection of Winsor I phase behavior (lower phase microemulsion) at low salinity yields good recovery and low surfactant retention. Also, intermediate-water conditions seem to be more favorable compared to water-wet conditions. Moreover, surfactant injection with low salinity water gives higher oil recovery than the expected performance using only surfactant flooding.

Moreover, Alagic and Skauge (2010) investigated the combined effect of injecting low salinity water with surfactant flooding on oil recovery from Berea sandstones. The authors conducted several corefloods and highlighted the positive response obtained using the combined injection. In a later work, Alagic et al. (2011) confirmed their previous findings on longer cores to minimize capillary-end effects and investigated the effect of aging on recovery. They reported higher oil recovery using the combined injection from aged cores compared to unaged cores. They reported that the combined low salinity and surfactant floodings prevents retrapping of the mobilized oil by LSWI. Tavassoli et al. (2015) used the UTCHEM-IPHREEQC simulator to investigate the combined effect of LSWI and surfactant flooding (UTCHEM Technical Documentation, 2000). The authors history-matched the experimental work of Alagic and Skauge (2010) including oil recovery, effluent ionic composition, and pressure gradient data. Moreover, they performed several simulations from which they concluded that the high-salinity surfactant flood performs better than the low salinity surfactant flood. They stressed on the importance of surfactant selection and design rather than the benefits of low salinity on a surfactant flood.

Also, Khanamiri et al. (2015) studied the performance of surfactant injection when low salinity surfactant is combined with LSWI, as well as the combination of near optimal salinity surfactant with LSWI. The authors conducted several corefloods on Berea sandstone cores. They concluded that tertiary injection of low salinity surfactant after a secondary LSWI outperforms after-tertiary low salinity surfactant when high-salinity and low salinity waters are injected in secondary and tertiary modes, respectively. Also, they found that near-optimal salinity surfactant performs better after a secondary injection of a high-salinity water compared to a low salinity water; however, the latter performs better if it is followed with a high-salinity chase-water injection.

7.5 LSWI/EWI AND CO_2 FLOODING APPLICATION

SWAG and WAG are the two proposed injection forms for gas mobility control. The idea of having a mixture of CO_2 and water, whether in WAG or SWAG mode, is to contact the bypassed oil after the CO_2 slug. Hence, the properties of the injected water play a vital role in affecting the level of incremental oil recovery (Aleidan and Mamora, 2010). Several factors may affect CO_2-WAG including reservoir heterogeneity, fluid properties, miscibility condition, rock wettability, and WAG parameters such as water–gas slug size, timing of injection, and WAG ratio (Jiang et al., 2010). The main experimental and numerical works conducted for investigating the combined effect of low salinity water and CO_2 flooding is discussed.

Kulkarni and Rao (2005) conducted miscible and immiscible WAG with varying brine composition on Berea sandstone cores. They reported a decrease in oil recovery with decreasing the salinity of the injected water due to the increase in the solubility of CO_2 in brine. Jiang et al. (2010) investigated the effect of injection brine salinity on CO_2-WAG performance in the tertiary mode on Berea sandstone cores by changing the salinity of the injected brine systematically up to 32,000 ppm. The coreflooding experiments were run at 60°C and at a pressure 20% above the minimum miscibility pressure (MMP) to ensure miscible CO_2 flooding. The results revealed that the WAG recovery increases with increasing the salinity of the injection brine; this was explained due to the salting-out effect, as the solubility of CO_2 in water decreases with increasing the salinity. This allows more CO_2 available for oil displacement resulting in higher oil recovery.

Other than CO_2 solubility, wettability alteration is another important factor for incremental oil recovery by combined low salinity and CO_2 injections. Fjelde and Asen (2010) investigated wettability alteration during water flooding and CO_2 flooding on reservoir chalk rocks from the North Sea at different temperatures (50°C and 130°C). The experimental work started with formation water as a first phase, followed by seawater as a second phase, and finally a cycle of seawater alternating carbon dioxide flooding (CO_2-WAG) as a third phase. The results showed that wettability alteration towards more water-wet was observed after the WAG slug resulting in residual oil saturation between 3% and 5%.

On the other hand, the combined effect of injection on carbonates was studied by Aleidan and Mamora (2010). They studied the effect of different CO_2 injection modes on oil recovery by conducting coreflooding experiments including continuous gas injection (CGI), water alternating gas (WAG), and simultaneous water alternating gas injection (SWAG). The experiments were conducted at 120°F and 1900 psi, which is 100 psi more than the MMP using outcrop limestone carbonate cores. The salinity of the injected water was varied between 0 wt%, 6 wt%, and 20 wt%. The results showed that waterflooding recovery alone was not affected by salinity; this indicates that wettability alteration is not affecting oil recovery and the only controlling parameter is CO_2 solubility in water. SWAG and WAG resulted in higher oil recovery than CGI due to mobility control of the displacement front. SWAG resulted in the highest oil recovery and lowest CO_2 requirements. For both SWAG and WAG, decreasing salinity level leads to increasing CO_2 solubility in water and increasing oil recovery.

Moreover, Teklu et al. (2014) conducted several corefloods on carbonate and sandstone cores to investigate the combined effects of injecting low salinity water and CO_2 on oil recovery. Consecutive injections of seawater, low salinity water, and continuous CO_2 resulted in further reduction in the residual oil saturation. Contact angle and IFT were measured for the different rock samples and fluids to justify the underlying mechanism. Results showed a further decrease in both contact angle and IFT using the combined CO_2 and low salinity water. The authors believe that wettability alteration is further enhanced using the combined LSWI and CO_2. Reducing the salinity of injected water increases the solubility of CO_2, which deceases the brine-CO_2 IFT. The latter even further decreases the CO_2-saturated-brine and oil IFT and alters wettability towards a more water-wet state.

From a simulation point of view, Dang et al. (2014) presented a detailed evaluation of CO_2-LSWAG from a one-dimensional heterogeneous model into full-field simulation. They highlighted through simulations the combined benefits of gas and low salinity water floods by including geochemical reactions associated with CO_2 injection, ion-exchange process, and wettability alteration. A scaled ion-exchange equivalent fraction based on calcium adsorption on clay was used to interpolate between two sets of relative permeability representing water-wet and oil-wet systems. The authors concluded that CO_2-LSWAG overcomes the late production problem frequently encountered in conventional WAG processes. CO_2-LSWAG resulted in incremental oil recovery of 4.5%–9% OOIP. They reported that the success of CO_2-LSWAG depends on clay type and quantity, initial reservoir wettability condition, reservoir heterogeneity, reservoir minerals such as calcite and dolomite, composition of formation and injected brines, reservoir pressure and temperature for achieving CO_2 miscible condition, and WAG parameters.

Al-Shalabi et al. (2014a) used geochemical modeling to study the combined effect of LSWI and CO_2 on carbonate oil reservoirs. The most abundant acid in natural water systems responsible for rock weathering is carbonic acid. The carbonic acid forms as a result of CO_2 dissolution in water upon which $CO_{2(g)}$ becomes aqueous $CO_{2(aq)}$ and associates with water molecules as follows (Langmuir, 1997):

$$CO_{2(g)} \rightarrow CO_{2(aq)}, \quad (7.i.)$$

$$CO_{2(aq)} + H_2O \leftrightarrow H_2CO_3. \quad (7.ii)$$

By convention, the two species of $CO_{2(aq)}$ and H_2CO_3 are summed up as $H_2CO_3^*$ which is given by

$$CO_{2(g)} + H_2O \leftrightarrow H_2CO_3^*. \quad (7.iii)$$

Then, the formed carbonic acid affects both calcite and dolomite directly, and anhydrite indirectly depending on the pH of the formation. The first stage of carbonic acid dissociation is as the following (Appelo and Postma, 2010):

$$H_2CO_3^* \leftrightarrow H^+ + HCO_3^-. \quad (7.iv)$$

The second stage of dissociation is given by

$$HCO_3^- \leftrightarrow H^+ + CO_3^{2-}. \quad (7.v)$$

For calcite, the equation of calcite dissolution is

$$CaCO_3 \leftrightarrow Ca^{2+} + CO_3^{2-}. \quad (7.vi)$$

By adding Eqs. (7.iii–7.vi), the effect of CO_2 injection on calcite can be summarized as follows:

$$CO_{2(g)} + H_2O + CaCO_3 \leftrightarrow Ca^{2+} + 2HCO_3^-. \quad (7.vii)$$

Hence, more dissolution of calcite occurs with CO_2 injection. Similarly, the dissolution reaction for dolomite can be written as follows:

$$2CO_{2(g)} + 2H_2O + CaMg(CO_3)_2 \leftrightarrow Ca^{2+} + Mg^{2+} + 4HCO_3^-. \quad (7.viii)$$

For anhydrite, the effect of CO_2 is considered indirectly through the changes in calcium ion, which affects the dissolution of anhydrites given by

$$CaSO_4 \leftrightarrow Ca^{2+} + SO_4^{2-}. \quad (7.ix)$$

Nevertheless, by including the LSWI effect on the reactions in the presence of CO_2 along with different conditions of temperature, pressure, and pH, the trends for dissolution/precipitation of calcite, dolomite, and anhydrite change.

The geochemical modeling of the coreflood of Yousef et al. (2011) was performed by Al-Shalabi et al. (2014a) using the PHREEQC simulator (Parkhurst and Appelo, 2013). Three injection modes were compared including LSWI only, CO_2 injection only, and combined LSWI and CO_2 injection. Geochemical analysis showed that the pH-induced wettability alteration is more pronounced using LSWI only than the combined effect of LSWI and CO_2 injection (Fig. 7.2). This is due to the increase in pH trend in LSWI only compared to the decrease in the same trend using CO_2 only and the combined LSWI and CO_2 injection. The latter decrease in pH can be best explained by the formation of carbonic acid.

Moreover, the combined effect of LSWI and CO_2 is most prominent on carbonates with high dolomite composition (Fig. 7.3), whereas carbonates with high anhydrite composition are most affected by LSWI only (Fig. 7.4).

We cannot generalize the findings of the previous study, as the combined effect of LSWI and CO_2 injection on oil recovery is case dependent based on temperature, pressure, rock lithology, oil type, initial rock wettability state, and injected water composition.

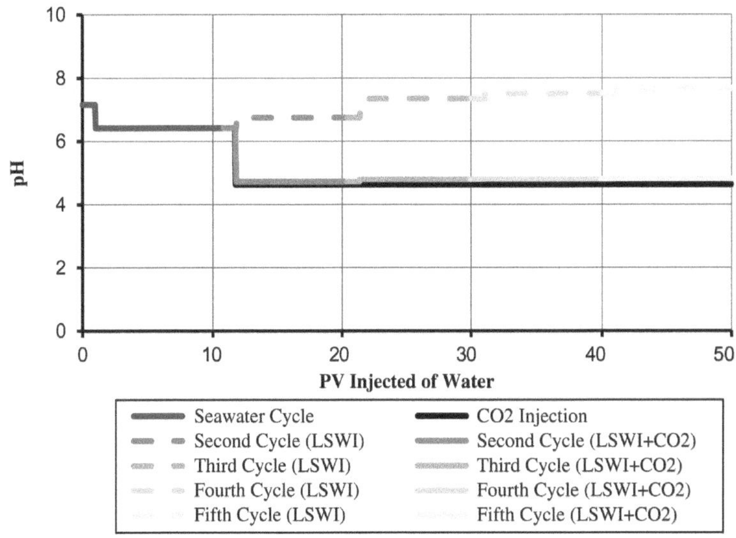

Figure 7.2 pH number using PHREEQC-fluid species (LSWI, CO_2, and LSWI+CO_2) (Al-Shalabi et al., 2014a).

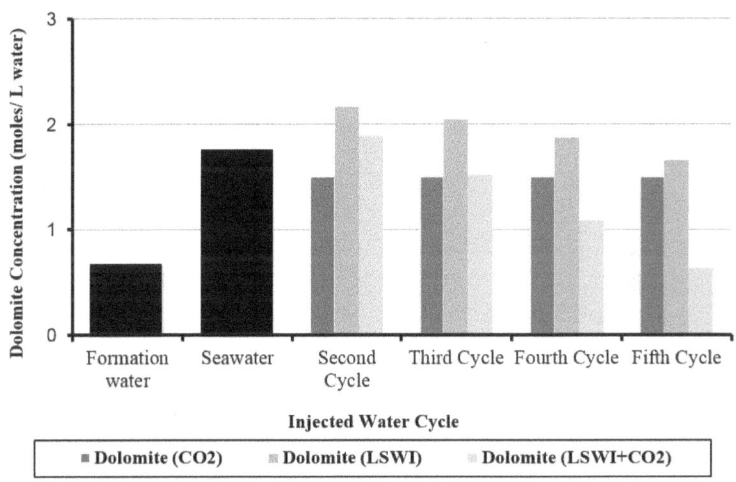

Figure 7.3 Dolomite concentration at different injection cycles (LSWI, CO_2, and LSWI+CO_2) (Al-Shalabi et al., 2014a).

Later, Al-Shalabi et al. (2016) investigated the combined effect of injecting low salinity water (LSWI) and carbon dioxide (CO_2) on oil recovery from carbonate cores using the UTCOMP simulator (UTCOMP Technical Documentation, 2003). They used 1D simulations as well as experimental corefloods for verifying the results. The simulation

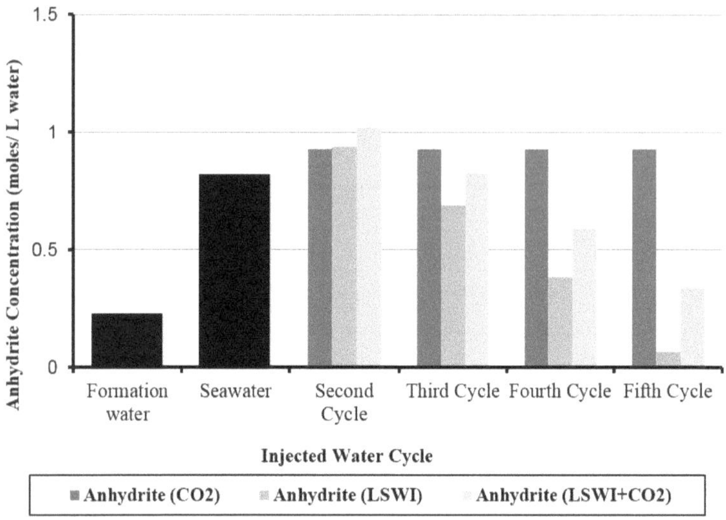

Figure 7.4 Anhydrite concentration at different injection cycles (LSWI, CO_2, and LSWI+CO_2) (Al-Shalabi et al., 2014a).

work includes miscible and immiscible CGI, SWAG, WAG, and tapered (WAG). Different dilutions of seawater were simulated. The CO_2 injected was above its MMP. The authors modified Baker's three-phase relative permeability model to account for the effect of salinity on the two-phase water/oil relative permeability as follows Baker (1988):

– Two-phase oil/water relative permeability

$$k_{row} = k^*_{row}\left(\frac{S_o - S_{orw}}{1 - S_{wirr} - S_{orw}}\right)^{eow}, \tag{7.1}$$

where e_{ow} is the oil/water exponent. Now, we add the low salinity water injection effect on oil/water endpoint relative permeability (k^*_{row}), oil/water exponent (e_{ow}), and residual oil saturation to water (S_{orw}) using the Empirical LSWI Model (Al-Shalabi et al., 2014b) as follows:

– LSWI effect on oil/water endpoint relative permeability

$$k^*_{row} = \frac{k^{*LS}_{row} - k^{*HS}_{row}}{1 + \left(\frac{\theta}{a}\right)^e} + k^{*HS}_{row}, \tag{7.2}$$

– LSWI effect on oil/water exponent

$$e_{ow} = \frac{e_{ow\max} - e^{LS}_{ow}}{1 + \left(\frac{\theta}{a}\right)^{-e}} + e^{LS}_{ow}, \tag{7.3}$$

– LSWI effect on residual oil saturation to water

$$S_{orw(LSWI)} = \omega \times S_{orw}^{LS} + (1 - \omega) \times S_{orw}^{HS}, \tag{7.4}$$

$$\omega = \frac{(\theta - \theta^{HS})}{(\theta^{LS} - \theta^{HS})}, \tag{7.5}$$

then a minimum value is selected for the residual oil saturation to water as follows:

$$S_{orw} = \min[S_o, S_{orw}, S_{orw(\mathrm{LSWI})}]. \tag{7.6}$$

The results showed that SWAG, whether using seawater or its dilutions, outperformed all other tertiary injection modes in terms of oil recovery. Moreover, the study highlighted the advantage of using LSWI along with miscible CO_2; miscible CO_2 displaces the residual oil saturation, whereas the LSWI boosts the production rate by increasing the oil relative permeability through wettability alteration towards more water-wet state (Fig. 7.5).

The latter finding was supported by Chandrasekhar and Mohanty (2014) corefloods, which were conducted in SWAG tertiary mode using seawater and its dilutions (Fig. 7.6).

In addition, the study showed the advantages of using SWAG with low salinity water from a fractional flow analysis. Fractional flow analysis was performed on SWAG (SW, LSWI) and miscible CGI processes by

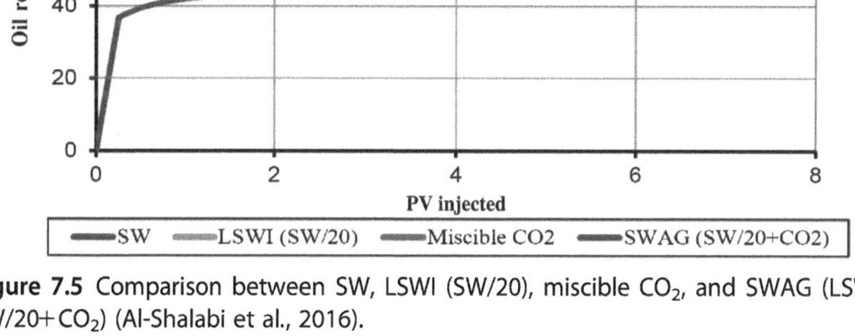

Figure 7.5 Comparison between SW, LSWI (SW/20), miscible CO_2, and SWAG (LSWI-SW/20+CO_2) (Al-Shalabi et al., 2016).

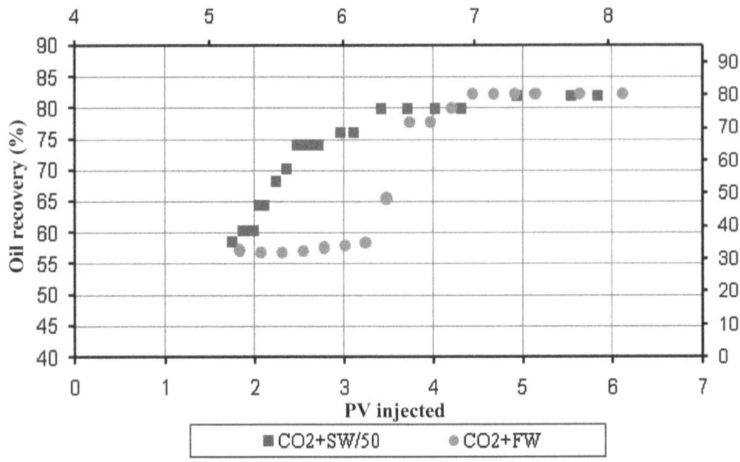

Figure 7.6 Cumulative oil recovery for SWAG experiments in the tertiary mode (Chandrasekhar and Mohanty, 2014).

following the approach suggested by Walsh and Lake (1988) and including the low salinity water effect on the water–solvent curve. The main assumptions made for the fractional flow analysis include neglecting the followings:

- Gravity effects
- Capillary effects
- Solvent–water solubility
- Trapped oil saturation
- Solvent partitioning into the trapped oil saturation

The resulting fractional flow curves for water–oil and water–solvent are as follows:

– Water/oil fractional flow curve:

$$f_{w/o} = \frac{1}{1 + \dfrac{k_{ro}}{k_{rw}} \dfrac{\mu_w}{\mu_o}}, \tag{7.7}$$

– Water/solvent fractional flow curve:

$$f_{w/s} = \frac{1}{1 + \dfrac{k_{ro}}{k_{rw}} \dfrac{\mu_w}{\mu_s}}. \tag{7.8}$$

For SWAG (SW + CO_2) and CGI processes, the water/solvent fractional flow curve is obtained from the water/oil fractional flow curve by simply replacing the oil viscosity by the CO_2 gas viscosity. For SWAG (LSWI + CO_2), a step further should be taken by including the effect of LSWI on oil relative permeability and water viscosity in Eq. (7.8). The relative permeability parameters used for constructing fractional flow curves for SWAG (SW, LSWI) and miscible CGI are presented in Table 7.1.

The fractional flow curves along with total mobility for SWAG (SW, LSWI) and miscible CGI are shown in Figs. 7.7–7.12. The total mobility

Table 7.1 Relative permeability parameters used for constructing fractional flow curves for SWAG (SW, LSWI) and miscible CGI processes

Tertiary injection mode	Miscible CGI/ SWAG ($SW+CO_2$)		SWAG ($LSWI+CO_2$)	
Parameter	**Water–oil**	**Water–solvent**	**Water–oil**	**Water–solvent**
k^*_{rw}	0.025	0.025	0.025	0.025
k^*_{ro}	0.203	0.203	0.203	**0.96**
n_w	1.3	1.3	1.3	1.3
n_o	3.5	3.5	3.5	**1.53**
S_{wirr}	0.3181	0.3181	0.3181	0.3181
S_{orw}	0.329	0.329	0.329	0.329

Source: From Al-Shalabi, E.W., Sepehrnoori, K., Pope, G., 2016. Numerical modeling of combined low salinity water and carbon dioxide in carbonate cores. J. Petrol. Sci. Eng. 137, 157–171.

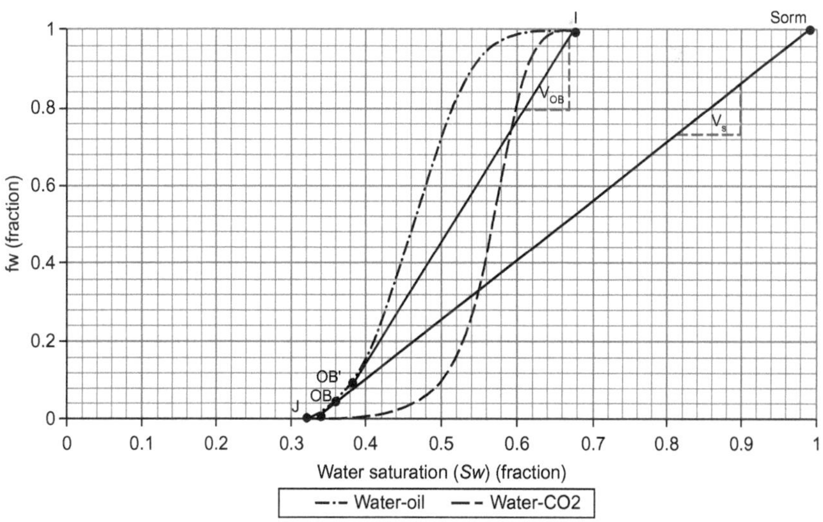

Figure 7.7 Fractional flow curves for miscible CGI (Al-Shalabi et al., 2016).

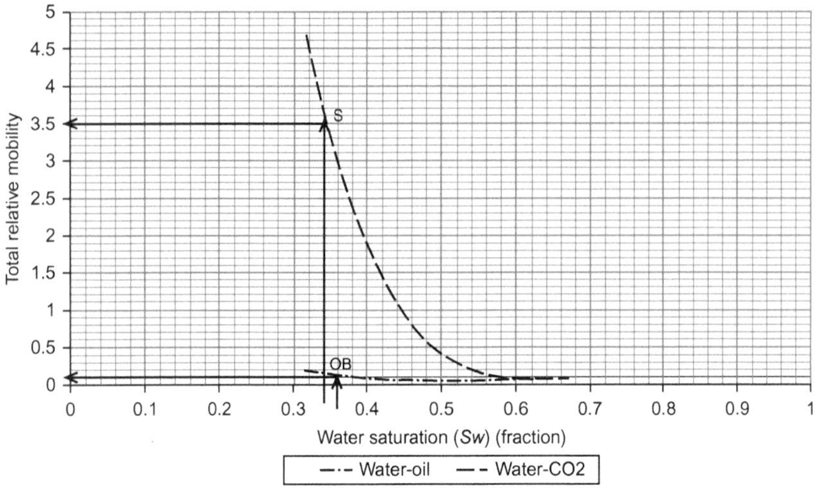

Figure 7.8 Total mobility curves for miscible CGI (Al-Shalabi et al., 2016).

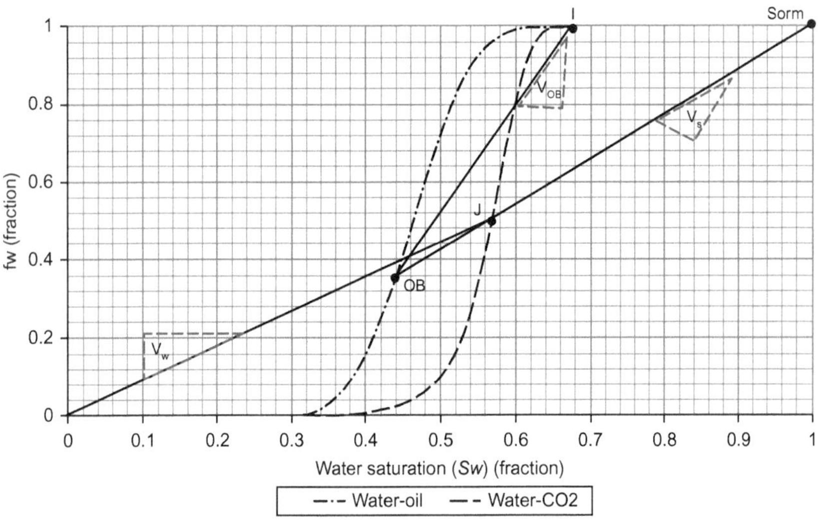

Figure 7.9 Fractional flow curves for SWAG (SW+CO_2) (Al-Shalabi et al., 2016).

curves were used to calculate the effective displacement mobility ratio as follows:

$$M = \frac{\left[\left(k_{ro}/\mu_s\right) + \left(k_{rw}/\mu_w\right)\right]_{\text{solvent}}}{\left[\left(k_{ro}/\mu_o\right) + \left(k_{rw}/\mu_w\right)\right]_{\text{oil bank}}}. \tag{7.9}$$

The effective displacement mobility ratio gives a rough estimate of the stability of each process at field scale. Moreover, the optimum solvent usage

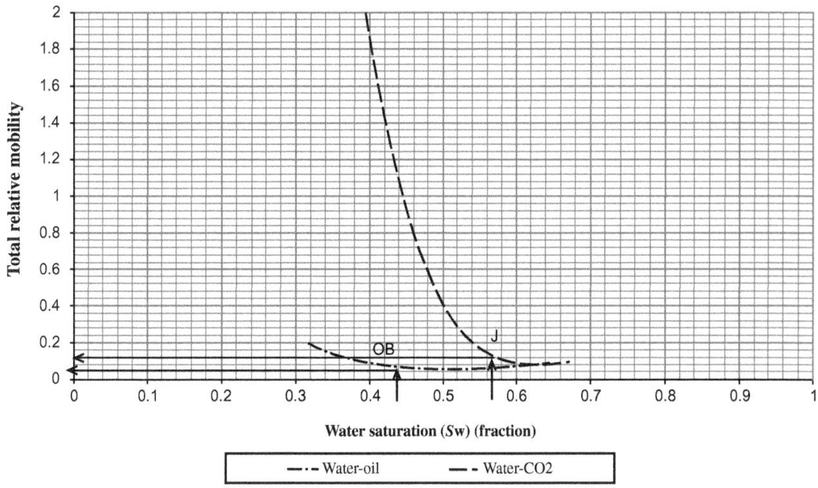

Figure 7.10 Total mobility curves for SWAG (SW+CO_2) (Al-Shalabi et al., 2016).

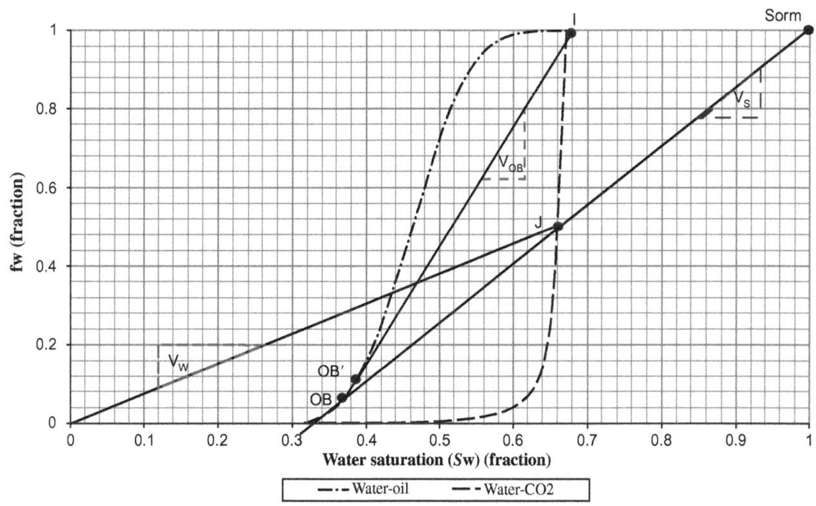

Figure 7.11 Fractional flow curves for SWAG (LSWI+CO_2) (Al-Shalabi et al., 2016).

was calculated for the sake of comparison between SWAG (SW, LSWI) and miscible CGI processes. The optimum solvent usage is given by

$$m_s = (1 - f_{wJ})t_{Ds}, \tag{7.10}$$

$$f_{wJ} = \frac{W_R}{1 + W_R}, \tag{7.11}$$

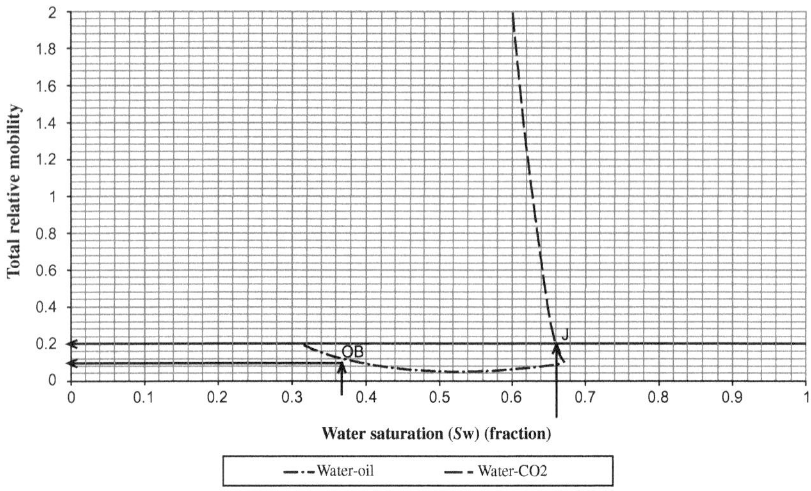

Figure 7.12 Total mobility curves for SWAG (LSWI+CO_2) (Al-Shalabi et al., 2016).

Table 7.2 M and m_s calculations for SWAG (SW, LSWI) and miscible CGI processes

Tertiary injection mode	Miscible CGI	SWAG (SW+CO_2)	SWAG (LSWI+CO_2)
$\lambda_{T\text{solvent}}$	3.5	0.12	0.2
$\lambda_{T\text{oil bank}}$	0.1	0.04	0.1
M	**35**	**3**	**2**
W_R	0	1	1
f_{wJ}	0	0.5	0.5
v_{cs}	1.53	1.15	1.50
m_s (PV)	**0.653**	**0.436**	**0.334**

Source: From Al-Shalabi, E.W., Sepehrnoori, K., Pope, G., 2016. Numerical modeling of combined low salinity water and carbon dioxide in carbonate cores. J. Petrol. Sci. Eng. 137, 157–171.

$$t_{Ds} \cong \frac{1}{v_{Cs}}, \tag{7.12}$$

where W_R is the simultaneous water–solvent ratio and v_{Cs} is concentration velocity of solvent.

The results for calculating effective displacement mobility ratio and optimum solvent usage for the three tertiary injection modes are listed in Table 7.2. The effective displacement mobility ratio was calculated right before and after solvent miscible displacing front, which is given by points of J and OB, respectively (Figs. 7.9 and 7.11), except for the miscible CGI case which is given by points S and OB (Fig. 7.7). The reason for that is to avoid intersecting the water–oil curve twice, where a tangent was drawn to the water–oil curve at point S. Hence, in this case we have a spreading wave from point J to S. More details about constructing fractional flow curves for simultaneous miscible gas-oil processes in the tertiary mode are described elsewhere (Walsh and Lake, 1988).

From a fractional flow point of view, Table 7.2 shows that SWAG (LSWI+CO_2) outperforms SWAG (SW+CO_2) and miscible CGI by achieving the lowest effective mobility ratio of 2 and the lowest optimum solvent usage of 0.334 PV. Also, the analysis shows that the miscible CGI is severely unstable and viscous fingering perturbations are highly expected due to the high mobility ratio of 35. This was expected as there is no water injection for improving the mobility control of the solvent slug.

Moreover, according to Walsh and Lake (1988), the optimal W_R occurs when the miscible wave line intersects the water–solvent fractional flow curve and is tangential to the water–oil curve. Based on that, the SWAG (SW+CO_2) is above the optimum (W_R=1) with a small oil fractional flow in the oil bank and large time for complete recovery. Nevertheless, for the same W_R of 1 used for SWAG (LSWI+CO_2), the process achieves the optimal W_R where the miscible wave line is tangential to the water–oil curve. This is basically due to the effect of LSWI on the water–solvent curve, which leads to achieve optimal condition at similar injected W_R. Hence, the use of low salinity water reduces the amount of solvent required to achieve optimal conditions, which results in large oil fractional flow, quick recovery, and even a more profitable processes with better economics.

After discussing the different applications of low salinity/engineered water injection from both laboratory and simulation aspects, a comparison of the main controlling factors for the effect of low salinity/engineered water injection on sandstones and carbonates is given in the next chapter.

REFERENCES

Alagic, E., Skauge, A., 2010. Combined low salinity brine injection and surfactant flooding in mixed-wet sandstone cores. Energ. Fuel. 24, 3551–3559.

Alagic, E., Spildo, K., Skauge, A., Solbakken, J., 2011. Effect of crude oil ageing on low salinity and low salinity surfactant flooding. J. Petrol. Sci. Eng. 78 (2), 220–227.

Aleidan, A., Mamora, D.D., 2010. SWACO_2 and WACO_2 efficiency improvement in carbonate cores by lowering water salinity. Canadian Unconventional Resources and International Petroleum Conference, Calgary, Alberta, Canada, Paper SPE 137548.

Al-Shalabi, E.W., Sepehrnoori, K., Pope, G., 2014a. Geochemical investigation of the combined effect of injecting low salinity water and carbon dioxide on carbonate

reservoirs. International Conference on Greenhouse Gas Technologies (GHGT), Austin, Texas, USA, Paper GHGT 1202 (63), 7663–7676.
Al-Shalabi, E.W., Sepehrnoori, K., Delshad, M., Pope, G., 2014b. A novel method to model low-salinity-water injection in carbonate oil reservoirs. SPE J. 20 (5), 1154–1166.
Al-Shalabi, E.W., Sepehrnoori, K., Pope, G., 2016. Numerical modeling of combined low salinity water and carbon dioxide in carbonate cores. J. Petrol. Sci. Eng. 137, 157–171.
Alzayer, H., Sohrabi, M., 2013. Numerical simulation of improved heavy oil recovery by low-salinity water injection and polymer flooding. SPE Annual Technical Symposium and Exhibition, Khobar, Saudi Arabia, Paper SPE 165287.
Appelo, C.A.J., Postma, D., 2010. Geochemistry, Groundwater and Pollution. Second ed. Taylor & Francis Group Plc, Boca Raton, FL.
Ayirala, S., Ernesto, U., Matzakos, A., Chin, R., Doe, P., Hoek, P.V.D., 2010. A designer water process for offshore low salinity and polymer flooding applications. SPE Improved Oil Recovery Symposium, Tulsa, Oklahoma, USA, Paper SPE 129926.
Ayirala, S., Yousef, A., 2015. A state-of-the-art review to develop injection-water-chemistry requirement guidelines for IOR/EOR projects. SPE Prod. Oper. 30 (1), 26–42.
Baker, L., 1988. Three-phase relative permeability correlations. SPE Enhanced Oil Recovery, Tulsa, Oklahoma, USA, Paper SPE 17369.
Chandrasekhar, S., Mohanty, K.K., 2014. Private Communication.
Dang, C.T.Q., Nghiem, L.X., Chen, Z., Nguyen, N.T.B., Nguyen, Q.P., 2014. CO_2 low salinity water alternating gas: A New Promising Approach for Enhanced Oil Recovery. SPE Improved Oil Recovery Symposium, Tulsa, OK, USA, Paper SPE 169071.
Fjelde, I., Asen, S.M., 2010. Wettability alteration during water flooding and carbon dioxide flooding of reservoir chalk rocks. SPE EUROPEC/EAGE Annual Conference and Exhibition, Barcelona, Spain, Paper SPE 130992.
Han, B., Lee, J., 2014. Sensitivity analysis on the design parameters of enhanced oil recovery by polymer flooding with low salinity waterflooding. International Society of Offshore and Polar Engineers, The Twenty-fourth International Ocean and Polar Engineering Conference, Busan, Korea, Paper SPE 130992.
Jiang, H., Nuryaningsih, L., Adidharma, H., 2010. The effect of salinity of injection brine on water alternating gas performance in tertiary miscible carbonate dioxide flooding: Experimental Study. SPE Western Regional Meeting, California, USA, Paper SPE 132369.
Jose, S.R., Gachuz-Muro, H., Sohrabi, M., 2015. Application of low salinity water injection in heavy oil carbonate. SPE EUROPEC, Madrid, Spain, Paper SPE 174391.
Khanamiri, H.H., Torsaeter, O., Stensen, J.A., 2015. Experimental study of low salinity and optimal salinity surfactant injection. SPE EUROPEC, Madrid, Spain, Paper SPE 174367.
Kozaki, C., 2012. Efficiency of low salinity polymer flooding in sandstone cores. Master's Thesis, The University of Texas at Austin, Texas, USA.
Kulkarni, M.M., Rao, D.N., 2005. Experimental investigation of miscible and immiscible water-alternation-gas (WAG) process performance. J. Petrol. Sci. Eng. 48 (1), 1–20.
Langmuir, D., 1997. Aqueous Environmental Geochemistry. Prentice-Hall, Inc, Upper Saddle River, NJ.
Mohammadi, H., Jerauld, G.R., 2012. Mechanistic modeling of the benefit of combining polymer with low salinity water for enhanced oil recovery. SPE Improved Oil Recovery Symposium, Tulsa, Oklahoma, USA, Paper SPE 153161.
Parkhurst, D.L., Appelo, C.A.J., 2013. Description of Input and Examples for PHREEQC Version 3—A Computer Program for Speciation, Batch-Reaction,

One-Dimensional Transport, and Inverse Geochemical Calculations. Chapter 43 of Section A Groundwater, Book 6 Modeling Techniques.

Spildo, K. Johannessen, A.M., Skauge, A., 2012. Low salinity waterflood at reduced capillary. SPE Improved Oil Recovery Symposium, Tulsa, Oklahoma, USA, Paper SPE 154236.

Tavassoli, S., Korrani, A.K.N., Pope, G.A., Sepehrnoori, K., 2015. Low salinity surfactant flooding-a multi-mechanistic enhanced oil recovery method. SPE International Symposium on Oilfield Chemistry, The Woodlands, Texas, USA, Paper SPE 173801.

Teklu, T.W., Alameri, W., Graves, R.M., Kazemi, H., AlSumaiti, A.M., 2014. Low-salinity water-alternating-CO_2 flooding enhanced oil recovery: Theory and Experiments. Abu Dhabi International Petroleum Exhibition and Conference, Abu Dhabi, UAE, Paper SPE 171767.

UTCHEM−9.0 Technical Documentation, 2000. The University of Texas at Austin, Volume II, Texas, USA.

UTCOMP − 3.8 Technical Documentation, 2003. The University of Texas at Austin, Texas, USA.

Vermolen, E.C.M., Pingo-Almada, M., Wassing, B.M., Ligthelm, D.J., Masalmeh, S.K., 2014. Low-salinity polymer flooding: Improving polymer flooding technical feasibility and economics by using low-salinity make-up brine. SPE International Petroleum Technology conference, Doha, Qatar, Paper SPE 17342.

Walsh, M.P., Lake, L.W., 1988. Applying fractional flow theory to solvent flooding and chase fluids. J. Petrol. Sci. Eng. 2, 281−303.

CHAPTER EIGHT

Comparison of LSWI/EWI Effect on Sandstone and Carbonate Rocks

Content

This chapter highlights the main differences in the applicability of low salinity/engineered water flooding techniques on carbonates compared to sandstones. Almost all producing sandstone rocks contain clays as a coating on individual sand grains and/or discrete particles mixed with the sand. Carbonates may contain clays as well, but these clays are usually encapsulated in the rock matrix and not significantly affected by the invading fluids. Sands that contain clay from 1% to 5% are called clean sands, but dirty sand is used usually for describing sands with 5% to more than 20% clay. Clay types present in sandstone rocks are usually smectite, illite, mixed-layer clays (primarily illite-smectite), kaolinite, and chlorite (Alotaibi and Nasr-El-Din, 2009).

Low salinity/engineered water injection disturbs the initially established thermodynamic equilibrium between rock/oil/formation brine, which leads to a new equilibrium between the different phases resulting in favorable wettability alteration and increase of oil recovery during production period. In this process of wettability alteration, the activation energy is important as it controls the rate of chemical reactions between the mineral surface and injected water. If the reaction rate is too slow, there will be no improvement in wettability and oil recovery during the water injection period. Reservoir temperature plays a catalytic role in increasing the rate of chemical reactions, as there is a strong relation between activation energy and temperature as was reported by Puntervold et al. (2007).

The activation energy needed for wettability modification depends on both the strength of bonding between the polar oil components and the mineral surface, and the reactivity of the ions present in the injected

Low Salinity and Engineered Water Injection for Sandstone and Carbonate Reservoirs.
DOI: http://dx.doi.org/10.1016/B978-0-12-813604-1.00008-0

water. In general, the bonding energy between the negative polar crude oil components and carbonates are stronger than the one existing in clays and silicate (sandstone rocks) as reported by Thomas et al. (1993). Doust et al. (2009) stated that there is difference between the chemical bonding of crude oil negative polar components (carboxylic material) with positively charged carbonate rocks and negatively charged sandstone rocks, which promotes the difference in wettability alteration mechanism. Thus, removal of the organic material in carbonate rock is achieved by increasing the surface reactivity of the potential determining ions Ca^{2+}, Mg^{2+}, and SO_4^{2-} at high temperatures. Adsorption of organic material is essential for low salinity/ engineered water effect in sandstone rocks; however, desorption of these material at high temperature helps in having more pronounced effect for low salinity/engineered water injection (LSWI/EWI) on carbonate rocks.

In carbonates, seawater can cause the wettability alteration effect; however, low salinity water injection (<5000 ppm) is required in case of sandstone rocks. As stated by Lager et al. (2006), contrary to chalk rocks, it is difficult to have a reliable chemical reaction model, which illustrates the multi-ion exchange (MIE) mechanism in sandstone with decreasing salinity; there exists a net desorption of adsorbed ions from the clay surface into the aqueous phase. Doust et al. (2009) reported that salting-in effect does not work in carbonate rock due to strong bonding of organic matter, whereas it is effective in sandstone rocks due to the weak bonding of the organic material to the clay surface.

At low pH, crude oils are positively charged; as the pH value starts increasing, crude oil charge decreases to zero at the isoelectric point [point of zero charge (PZC)] and then becomes strongly negatively charged at high pH (Takamura and Chow, 1985). Sandstone rocks are negatively charged above pH of 2 (Menezes et al., 1989). The PZC for limestone is around 9.2,while the PZC for dolomite is about 7.4 (Gupta and Mohanty, 2010). The PZC for calcite and dolomite depends on pH and solution composition, where calcite PZC ranges from 7 to 12 while dolomite PZC ranges from 6 to 8.8 (Pokrovsky et al., 2002). Lichaa et al. (1992) reported PZC values of 4.6 and 3.4 for two carbonate rocks located in Saudi Arabia. The cores were tested in deionized water, and they compose of 40% calcite and 60% dolomite (Fig. 8.1). The PZC of dolomite is lower than that of limestone (Alotaibi et al., 2011).

Stability of water film depends on the electrical double layer repulsion, which results from the surface charge at both solid/water, and water/oil interfaces. In case both interfaces have similar charges, a repulsive

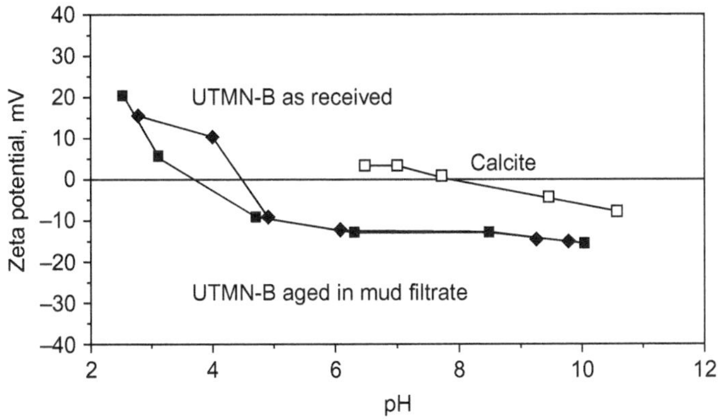

Figure 8.1 Variation of zeta potential with pH of two carbonate rocks and calcite tested in deionized water (Lichaa et al., 1992).

electrostatic force occurs maintaining high disjoining pressure and a thick water film forms, which results in a water-wet rock surface (Dubey and Doe, 1993). Lee et al. (2010) measured the water film thickness using sophisticated physical chemistry techniques such as the small angle scattering technique. They reported that the decrease in water salinity (ionic strength) causes increase in water layer thickness, which gives a more water-wet state for both sand and clay type particles. Moreover, a lower divalent cation concentration in the injected brine compared to the connate brine results in a larger low salinity improved oil recovery (IOR) response.

Ayirala and Yousef (2014) highlighted the importance of water chemistry for different IOR/enhanced oil recovery (EOR) processes, including offshore water flooding. Moreover, they defined SmartWater as a special type of water cocktail consisting of varying composition of different water ions. They stressed on the need for reducing the nonactive salts concentrations (Na^+ and Cl^-) as they prevent the accessibility of active potential determining ions to the rock surface. Also, they suggested using water salinity of less than 5000 ppm with a lower ionic strength and a small amounts for divalent cations (<500 ppm) to avoid clay swelling and hence improve oil recovery from sandstone rocks. On the other hand, low to moderate salinity diluted seawater (2–10-times dilution/28,000–6000 ppm) or modified seawater depleted in monovalent ions and enriched in potential determining divalent ions were recommended for better oil recovery by low salinity from carbonates.

In general, the mechanism behind wettability alteration of sandstone rocks using low salinity/engineered water injections is considered more complicated compared to carbonates due to the involved mechanisms with their various contributions including fine migration, pH increase, multiion exchange (MIE), and salting-in effects (Tang and Morrow, 1999; McGuire et al., 2005; Lager et al., 2006; Doust et al., 2009).

A comparison between carbonates and sandstones in terms of incremental oil recovery obtained by LSWI/EWI is presented in Table 8.1. The table shows a summary of the main LSWI/EWI corefloods conducted in both carbonate and sandstone rocks. Several parameters are listed in the table, which we believe affect incremental oil recovery. These parameters are rock type, temperature, pressure, salinity of injected and formation waters, oil viscosity, oil total acid number, and injection mode. Rock type matters, as previously discussed, even at the mineralogical level. Also, it is worth mentioning that the effect of low salinity water is more pronounced on field core plugs as compared to outcrop plugs. For temperature and pressure conditions, usually higher incremental oil recovery by low salinity is obtained at reservoir conditions for both rock types as opposed to ambient conditions. Hence, it is advisable to conduct experimental corefloods at reservoir conditions for more representative results. Sandstones and carbonates result in more incremental oil recovery when the difference in salinity between injection and formation brines is more pronounced. It should be highlighted that dilution of seawater is the common practice in sandstones (LSWI); however, tuning of injected water salinity is common in carbonates (EWI). For oil properties, usually lower oil viscosity and higher oil total acid number result in higher incremental oil recovery for both sandstones and carbonates. In terms of injection mode, both rock types are consistent as mostly higher incremental oil recovery is obtained in the secondary injection mode compared to the tertiary injection mode. This might be related to wettability alteration by expansion of the double electric layer, where the double layer functions better in the presence of a continuous oil film as opposed to the presence of discontinuous oil blobs, which is present in the tertiary injection mode.

Table 8.1 shows there are always exceptions for the trends discussed above. Therefore, the bottom line is that the improvement of oil recovery by LSWI/EWI in both sandstone and carbonate rocks is case-dependent because all the parameters discussed above matter and should be optimized to enhance the oil recovery. The main conclusions and recommendations based on this vast literature review and our experiences are presented in Chapter 9, Concluding Remarks.

Table 8.1 Summary of main LSWI/EWI corefloods in carbonates and sandstones (Al-Shalabi and Sepehrnoori, 2016)

Coreflood study	Rock type	Temperature (°C)	Pressure (psi)	Injected water salinity (ppm)	Formation water salinity (ppm)	Oil viscosity (cP)	Oil TAN (mg KOH/g oil)	Injection mode	Additional oil recovery (% OOIP)
LSWI/EWI in sandstones									
Reiter (1961)	Sandstone	15.56	14.70	3100	12,100	37.0 @ 25.56°C	-	Secondary	21.3
Bernard (1967)	Synthetic and berea sandstone	15.56	14.70	10,000–1000 NaCl	150,000 NaCl	—	-	Tertiary	2.63–6.98
Al-Mumen (1990)	Berea sandstone	90	1500	5500–200,000 NaCl	200,000 NaCl	2.48 @ SC	-	Secondary	8–10
Zhang et al. (2007)	Sandstone	75	250	1480 SW 1500 NaCl	29,690 FW	20.1 @ RC	1.46	Secondary Tertiary	29.2 7–14
Patil et al. (2008)	Sandstone	93	300	5500–50 SW	22,000 SW	—	-	Secondary	14–28
Webb et al. (2005)	Sandstone	-	-	1500 SW	250,000 FW	—	-	Secondary Tertiary	14 8–9
Agbalaka et al. (2009)	Berea sandstone	80	1500	10,000 NaCl	40,000 NaCl	8.24 @ SC	-	Secondary Tertiary	5–6 25–35
Rivet et al. (2010)	Berea and field sandstone	55	14.70	870–1140 SW	30,510 SW	7.93 @ RC	-	Secondary Tertiary	15 0
Gamage and Thyne (2011)	Berea and field sandstone	-	-	1% FW	FW	8.0–11.50 @ SC	0.074	Secondary Tertiary	10–22 2–6
Fjelde et al. (2012)	Sandstone	80	72.19	1054.96–105.49 SW	105,496 FW	1.5 @ RC	-	Secondary Tertiary	9 3
Suijkerbuijk et al. (2014)	Sandstone	87	14.70	-	FW	2.95 @ RC	> 0.01	Secondary Tertiary	6 5

(*Continued*)

Table 8.1 (Continued)

Coreflood study	Rock type	Temperature (°C)	Pressure (psi)	Injected water salinity (ppm)	Formation water salinity (ppm)	Oil viscosity (cP)	Oil TAN (mg KOH/g oil)	Injection mode	Additional oil recovery (% OOIP)
LSWI/EWI in carbonates									
Bagci et al. (2001)	Unconsolidated limestone	50	145	10,000–40,000 NaCl	–	46.5 @ RC	-	Secondary	14
				20,000 KCl					18.4
Yousef et al. (2011)	Carbonate	100	3000	28,835–576.7 SW	213,000 FW	1.26 @ RC	0.25	Tertiary	18
Gupta et al. (2011)	Dolomite and limestone	121.11	4000	33,484 SW 4xSO_4^{2-}	181,273 FW	1.13 @ RC	0.11	Tertiary	5–9
				33,375 SW					7–9
				29,970 SW BO_3^{3-}					15
				29,970 SW PO_4^{3-}					20
Zahid et al. (2012)	Carbonate chalk outcrop	90	72.52	28,835–2883 SW	213,734 FW	3.21 @ RC	0.96	Tertiary	15–20 0
Chandrasekhar and Mohanty (2013)	Limestone	120	50	2181–872 SW	179,700 FW	1 @ RC	2.45	Secondary Tertiary	40 32
Al-Attar et al. (2013)	Carbonate	25	100	1000–5000 SW	197,584–224,987 FW	3.08 @ SC	-	Secondary	21.5
Awolayo et al. (2014)	Carbonate	110	3000	43,000 SW 0.5–8 × SO_4^{2-}	261,210 FW	1.927 @ 70°C	-	Tertiary	10
Alameri et al. (2015)	Carbonate	90.56	1800	25,679–1027 SW	100,000 FW	3.0 @ RC	-	Tertiary	5–7

SC: Standard conditions of temperature (60°F) and pressure (1 atm).
RC: Reservoir conditions of temperature and pressure.
SW: Seawater.
FW: Formation water.

REFERENCES

Agbalaka, C.C., Dandekar, A.Y., Patil, S.L., Khataniar, S., Hemsath, J.R., 2009. Coreflooding studies to evaluate the impact of salinity and wettability on oil recovery efficiency. Transport Porous Med. 76 (1), 77–94.

Alameri, W., Teklu, T.W., Graves, R.M., Kazemi, H., AlSumaiti, A.M., 2015. Experimental and numerical modeling of low-salinity waterflood in a low permeability carbonate reservoir. SPE Western Regional Meeting, Garden Grove, California, USA, Paper SPE 174001.

Al-Attar, H.H., Mahmoud, M.Y., Zekri, A.Y., Almehaideb, R.A., Ghannam, M.T., 2013. Low salinity flooding in a selected carbonate reservoir: Experimental Approach. EAGE Annual Conference & Exhibition, London, United Kingdom, Paper SPE 164788.

Al-Mumen, A.A., 1990. The effect of injected water salinity on oil recovery. Master's Thesis, King Fahad University of Petroleum and Minerals, Dhahran, Saudi Arabia.

Al-Shalabi, E.W., Sepehrnoori, K., 2016. A comprehensive review of low salinity/engineered water injections and their applications in sandstone and carbonate rocks. J. Petroleum Sci. Eng. 139 (2016), 137–161.

Alotaibi, M.B., Nasr-El-Din, H.A., 2009. Chemistry of injection water and its impact on oil recovery in carbonate and clastic formations. SPE International Symposium on Oilfield Chemistry, The Woodlands, Texas, USA, Paper SPE 121565.

Alotaibi, M.B., Nasr-El-Din, H.A., Fletcher, J.J., 2011. Electrokinetics of limestone and dolomite rock particles. SPE Reserv. Eval. Eng. Paper SPE 148701. 14 (5), 594–603.

Awolayo, A., Sarma, H., AlSumaiti, A.M., 2014. A laboratory study of ionic effect of smart water for enhancing oil recovery in carbonate reservoirs. SPE EOR Conference at Oil and Gas West Asia, Muscat, Oman, Paper SPE 169662.

Ayirala, S.C., Yousef, A.A., 2014. Injection water chemistry requirement guidelines for IRO/EOR. SPE Improved Oil Recovery Symposium, Tulsa, USA, Paper SPE 169048.

Bagci, S., Kok, M.V., Turksoy, U., 2001. Effect of brine composition on oil recovery by waterflooding. J. Petrol. Sci. Technol. 19 (3–4), 359–372.

Bernard, G.G., 1967. Effect of floodwater salinity on recovery of oil from cores containing clays. SPE California Regional Meeting, Los Angeles, California, USA, Paper SPE 1725.

Chandrasekhar, S., Mohanty, K.K., 2013. Wettability alteration with brine composition in high temperature carbonate reservoirs. SPE Annual Technical Conference and Exhibition, New Orleans, Louisiana, USA, Paper SPE 166280.

Doust, A.R., Puntervold, T.P., Strand, S., Austad, T.A., 2009. Smart water as wettability modifier in carbonate and sandstone. 15th European Symposium on Improved Oil Recovery, Paris, France.

Dubey, S.T., Doe, P.H., 1993. Base number and wetting properties of crude oils. SPE Reservoir Eng. 8 (3), 195–200.

Fjelde, I., Asen, S.M., Omekeh, A., 2012. Low salinity water flooding experiments and interpretation by simulations. SPE Improved Oil Recovery Symposium, Tulsa, Oklahoma, USA, Paper SPE 154142.

Gamage, P., Thyne, G., 2011. Comparison of oil recovery by low salinity waterflooding in secondary and tertiary recovery modes. SPE Annual Technical Conference and Exhibition, Denver, Colorado, USA, Paper SPE 147375.

Gupta, R., Mohanty, K.K., 2010. Wettability alteration mechanism for oil recovery from fractured carbonate rocks. Transport Porous Med. 87 (2), 635–652.

Gupta, R., Smith, G.G., Hu, L., Willingham, T., Cascio, M.L., Shyeh, J.J., et al., 2011. Enhanced waterflood for middle east carbonates cores – Impact of injection water composition. SPE Middle East Oil and Gas Show and Conference, Manama, Bahrain, Paper SPE 142668.

Lager, A., Webb, K.J., Black, C.J.J., Singleton, M., Sorbie, K.S., 2006. Low salinity oil recovery- An experimental investigation. Proceedings of International Symposium of the Society of Core Analysts, Norway.

Lee, S.Y., Webb, K.J., Collins, I.R., Lager, A., Clarke, S.M., O'Sullivan, M., et al., 2010. Low salinity oil recovery-increasing understanding of the underlying mechanisms. SPE Symposium on Improved Oil Recovery, Tulsa, Oklahoma, USA, Paper SPE 129722.

Lichaa, P.M., Alpustun, H., Abdul, J.H., Nofal, W.A., Fuseni, A.B., 1992. Wettability evaluation of a carbonate reservoir rock. Advances in core evaluation III reservoir management, European Core Analysis Symposium, Paris, France, p. 327.

McGuire, P.L., Chatham, J.R., Paskvan, F.K., Sommer, D.M., Carini, F.H., 2005. Low salinity oil recovery: An Exciting New EOR Opportunity for Alaska's North Slope. SPE Western Regional Meeting, Irvine, California, USA, Paper SPE 93903.

Menezes, J.L., Yan, J., Sharma, M.M., 1989. The mechanism of wettability alteration due to surfactants in oil-based muds. SPE International Symposium on Oilfield Chemistry, Houston, Texas, USA, Paper SPE 18460.

Patil, S., Dandekar, A.Y., Patil, S.L., Khataniar, S., 2008. Low salinity brine injection for EOR on Alaska North Slope (ANS). International Petroleum Technology Conference, Kuala Lumpur, Malaysia, Paper SPE 12004.

Pokrovsky, O.S., Schott, J., Mielczarski, J.A., 2002. Surface speciation of dolomite and calcite in aqueous solutions, Encyclopedia of Surface and Colloid Science, 4. Marcel Dekker, New York, pp. 5081–5095.

Puntervold, T., Strand, S., Austad, T., 2007. Waterflooding of carbonate reservoirs: Effects of a Model Base and Natural Crude Oil Bases on Chalk Wettability. Energ. Fuel 21 (3), 1606–1616.

Reiter, Pl. K., 1961. A water-sensitive sandstone flood using low salinity water. Master's Thesis, University of Oklahoma, USA.

Rivet, S., Lake, L.W., and Pope, G.A., 2010. A coreflood investigation of low salinity enhanced oil recovery. SPE Annual Technical Conference and Exhibition, Florence, Italy, Paper SPE 134297.

Suijkerbuijk, B.M.J.M., Sorop, T.G., Parker, A.R., Masalmeh, S.K., Chmuzh, I.V., Karpan, V.M., et al., 2014. Low salinity waterflooding at West Salym: Laboratory Experiments and Field Forecasts. SPE EOR Conference at Oil and Gas West Asia, Muscat, Oman, Paper SPE 169691.

Takamura, K., Chow, R.S., 1985. The electric properties of the bitumen/water interface Part II. Application of the ionizable surface group model. Colloid. Surface 15 (1), 35–48.

Tang, G.Q., Morrow, N.R., 1999. Influence of brine composition and fines migration on crude oil/brine/rock interactions and oil recovery. J. Petrol. Sci. Eng. 24 (2–4), 99–111.

Thomas, M.M., Clouse, J.A., Longo, J.M., 1993. Adsorption of organic compounds on carbonate minerals. Chem. Geol. 109 (1–4), 227–237.

Webb, K.J., Black, C.J.J., Edmonds, I.J., 2005. Low Salinity Oil Recovery- The role of reservoir condition core floods. 13th European Symposium on Improved Oil Recovery, Budapest, Hungary.

Yousef, A.A., Al-Saleh, S., Al-Kaabi, A., Al-Jawfi, M., 2011. Laboratory investigation of the impact of injection-water salinity and ionic content on oil recovery from carbonate reservoirs. SPE Reserv. Eval. Eng. 14 (5), 578–593.

Zahid, A., Shapiro, A., Skauge, A., 2012. Experimental studies of low salinity water flooding in carbonate reservoirs: A New Promising Approach. SPE EOR Conference at Oil and Gas West Asia, Muscat, Oman, Paper SPE 155625.
Zhang, Y., Xie, X., Morrow, N.R., 2007. Waterflood performance by injection of brine with different salinity for reservoir cores. SPE Annual Technical Conference and Exhibition, Anaheim, California, USA, Paper SPE 109849.

CHAPTER NINE

Concluding Remarks

This book is a comprehensive state-of-the-art review on low salinity/engineered water injection for both sandstones and carbonates. The different aspects of low salinity/engineered water injection have been thoroughly discussed, including mechanisms, laboratory-, field-, and modeling-works as well as other applications. The following represent the recommendations and conclusions based on this vast literature review and our experiences:

- Wettability alteration is still believed to be the reason for the LSWI/EWI effect on oil recovery, especially from carbonates.
- Oil relative permeability parameters are more sensitive to LSWI compared to that of water phase.
- Electrical double layer (EDL) expansion might be contributing to the negligible change in the water relative permeability curves observed in some case studies as the immobile water film thickness increases and hence the relative permeability remains constant.
- Experiments are recommended to be conducted at reservoir conditions of temperature and pressure as well as using real reservoir cores and fluids for more representative results.
- Experimental results should be upscaled carefully to field scale as several pore volumes are usually used to obtain a certain incremental oil recovery by LSWI/EWI in the laboratory; however, this might mislead at field scale.
- Experimental-, field-, and numerical-works show the capability of LSWI/EWI on improving both displacement and volumetric sweep efficiencies.
- There is high potential of a hybrid LSWI/EWI technique with surfactant, polymer, and CO_2 floodings.
- The combined effect of LSWI/EWI is expected to be more pronounced at field scale, especially in cases of channeling and gravity override where the low salinity water contacts the unswept or bypassed zones.
- Geochemical modeling of the LSWI/EWI techniques is essential for better understanding of the complex reactions between rock/oil/brine systems.

Low Salinity and Engineered Water Injection for Sandstone and Carbonate Reservoirs.
DOI: http://dx.doi.org/10.1016/B978-0-12-813604-1.00009-2

- Modeling of LSWI/EWI with more emphasis on oil composition is recommended, especially for the cases of in situ surfactant generation where the acid number of crude oil becomes important.
- The improvement in oil recovery by LSWI/EWI in both sandstones and carbonates depends on temperature, pressure, rock mineralogy, oil type, initial rock wettability state, and injected water composition; so the results might vary.

For future work, research will progress in the area of low salinity/engineered water injection, including experimental, numerical, and field works. More work has to be conducted in order to understand the mechanism underlying incremental oil recovery by LSWI/EWI. The latter is important for optimizing the recipe of low salinity water as well as developing a reliable mechanistic model for predicting oil recovery at field scale. Techniques that are more complex can be used at nanoscale and integrated with micro- and macro-scales using a reliable simulator for good decision-making. Moreover, cheaper and less time-consuming techniques for screening and implementation of LSWI/EWI have to be investigated at both laboratory- and filed-scales under current oil market conditions. In addition, the synergistic effect of LSWI/EWI in combination with other EOR techniques has to be studied for boosting the low salinity effect, especially with the considerable pore volumes needed at laboratory scale.

APPENDIX A

Correlations Used for Estimating Fluid Properties

– Marhoun correlation (1988) to calculate gas solubility at reservoir conditions:

$$R_s = \left[a\gamma_g^b\gamma_o^c T^d P\right]^e$$

$$a = 185.843208$$

$$b = 1.877840$$

$$c = -3.1437$$

$$d = -1.32657$$

$$e = 1.398441$$

R_s is the gas solubility (SCF/STB)
P is the reservoir pressure (psi)
T is the reservoir temperature (°R)
γ_g is the specific gravity of solution gas
γ_o is the specific gravity of stock-tank oil

– Marhoun correlation (1988) to calculate oil formation volume factor at bubble-point pressure:

$$B_{ob} = 0.497069 + 0.862963 \times 10^{-3}T + 0.182594 \times 10^{-2}F + 0.318099 \times 10^{-5}F^2$$

$$F = R_s^a\gamma_g^b\gamma_o^c$$

$$a = 0.742390$$

$$b = 0.323294$$

$$c = -1.202040$$

R_s is the gas solubility (SCF/STB)
T is the reservoir temperature (°R)
γ_g is the specific gravity of solution gas
γ_o is the specific gravity of stock-tank oil

– Standing correlation (1981) to calculate oil density at the bubble-point pressure:

$$\rho_{ob} = \frac{62.4\gamma_o + 0.0136R_s\gamma_g}{0.972 + 0.000147\left[R_s\left(\frac{\gamma_g}{\gamma_o}\right)^{0.5} + 1.25(T-460)\right]^{1.175}}$$

ρ_{ob} is the bubble-point oil density (lb/ft^3)
T is the reservoir temperature (°R)

– Standing correlation (1981) to calculate the bubble-point pressure:

$$P_b = 18.2\left[\left(\frac{R_s}{\gamma_g}\right)^{0.83}\left(\frac{10^{0.00091(T-460)}}{10^{0.0125°API}}\right) - 1.4\right]$$

P_b is the bubble-point pressure (psi)
R_s is the gas solubility (SCF/STB)
API^o is the stock-tank oil API gravity
T is the reservoir temperature (°R)

Ahmed correlation (1989) to calculate the oil density at reservoir conditions:

$$\rho_o = \rho_{ob}\text{EXP}[B(\text{EXP}(aP) - \text{EXP}(aP_b))]$$
$$B = -(4.588893 + 0.0025999R_s)^{-1}$$
$$a = -0.00018473$$

ρ_o is the oil density (lb/ft^3)

– Ahmed correlation (1989) to calculate oil formation volume factor at reservoir conditions:

$$B_o = B_{ob}\text{EXP}\left[D\left[\text{EXP}(ap) - \text{EXP}(ap_b)\right]\right]$$
$$D = [4.588893 + 0.0025999R_s]^{-1}$$
$$a = -0.00018473$$

B_{ob} is the formation volume factor at saturation pressure (bbl/STB)
P is the reservoir pressure (psi)
P_b is the bubble-point pressure (psi)

– Glaso correlation (1980) to calculate the dead oil viscosity:

$$\mu_{od} = \left[3.141(10^{10})\right](T-460)^{-3.444}\left[\text{Log}(API)\right]^a$$
$$a = 10.313\left[\text{Log}(T-460)\right] - 36.447$$

μ_{od} is the dead oil viscosity (cP)
T is the reservoir temperature (°R)

– Standing correlation (1981) to calculate the bubble-point pressure oil viscosity:

$$\mu_{ob} = (10)^a(\mu_{od})^b$$

$$a = R_s\left[2.2(10^{-7})R_s - 7.4(10^{-4})\right]$$

$$b = \frac{0.68}{10^c} + \frac{0.25}{10^d} + \frac{0.062}{10^e}$$

$$c = 8.62(10^{-5})R_s$$

$$d = 1.1(10^{-3})R_s$$

$$e = 3.74(10^{-3})R_s$$

μ_{ob} is the bubble-point oil viscosity (cP)

– Standing correlation (1981) to calculate the oil viscosity at reservoir conditions:

$$\mu_o = \mu_{ob} + 0.001(P - P_b)(0.024\mu_{ob}{}^{1.6} + 0.038\mu_{ob}{}^{0.56})$$

μ_o is the oil viscosity (cP)
μ_{ob} is the bubble-point oil viscosity (cP)
P is the reservoir pressure (psi)
P_b is the bubble-point pressure (psi)

– Isdale and Morris correlation (1972) to calculate the seawater density at reservoir conditions:

$$\rho_{sw} = 10^3(A_1F_1 + A_2F_2 + A_3F_3 + A_4F_4)$$

$$B = (2S - 150)/150, \quad G_1 = 0.5, \quad G_2 = B, \quad G_3 = 2B^2 - 1$$

$$A_1 = 4.032G_1 + 0.115G_2 + 3.26 \times 10^{-4}G_3$$

$$A_2 = -0.108G_1 + 1.571 \times 10^{-3}G_2 - 4.23 \times 10^{-4}G_3$$

$$A_3 = -0.012G_1 + 1.74 \times 10^{-3}G_2 - 9 \times 10^{-6}G_3$$

$$A_4 = 6.92 \times 10^{-4}G_1 - 8.7 \times 10^{-5}G_2 - 5.3 \times 10^{-5}G_3$$

$$A = (2T - 200)/160, \; F_1 = 0.5, \; F_2 = A, \; F_3 = 2A^2 - 1, \; F_4 = 4A^3 - 3A$$

ρ_{sw} is the seawater density (kg/m^3)
T is the seawater temperature (°C)
S is the seawater salinity (ppm)

– Korosi and Fabuss correlation (1968) to calculate the seawater viscosity at reservoir conditions:

$$\mu_{sw} = \mu_w(1 + AS + BS^2)$$
$$A = 1.474 \times 10^{-3} + 1.5 \times 10^{-5}T - 3.927 \times 10^{-8}T^2$$
$$B = 1.073 \times 10^{-5} - 8.5 \times 10^{-8}T + 2.230 \times 10^{-10}T^2$$
$$\ln(\mu_w) = -0.00379418 + (0.604129/139.18 + T)$$

μ_{sw} is the seawater viscosity (kg/m s)
T is the seawater temperature (°C)
S is the seawater salinity (g/kg)

REFERENCES

Ahmed, T., 1989. Hydrocarbon Phase Behavior. Gulf Publishing Company, Houston.

Glaso, O., 1980. Generalized pressure–volume–temperature correlations. J. Pet. Technol. 32 (5), 785–795.

Isdale, J.D., Morris, R., 1972. Physical properties of seawater solutions: density. Desalination. 10 (4), 329–338.

Korosi, A., Fabuss, B.M., 1968. Viscosity of liquid water form 25°C to 150°C. J. Anal. Chem. 40 (1), 157–162.

Marhoun, M.A., 1988. PVT correlation for middle east crude oils. J. Pet. Technol. 40 (5), 650–660.

Standing, M.B., 1981. Volumetric and Phase Behavior of Oil Field Hydrocarbon Systems. 9th edition Society of Petroleum Engineers of AIME, Richardson, Texas.

APPENDIX B

Glossary

SYMBOLS

a	=	constant, inflection point from curve fitting
a_1, a_2	=	Langmuir adsorption parameters
C_a	=	mass fraction of salt component in aqueous phase
C_r	=	adsorption isotherm of salt onto rock
e	=	constant, hill slope from curve fitting
e_{ij}	=	exponent between i and j phases
e_{owmax}	=	maximum oil–water exponent
e_{ow}^{LS}	=	oil–water exponent when S_{or} becomes constant
F	=	scaling factor
I	=	total/stoichiometric ionic strength of solution
k	=	formation permeability
k_{rl}^{ow}	=	oil-wet phase relative permeability
k_{rl}^{ww}	=	water-wet phase relative permeability
k_{rl}^{*}	=	phase endpoint relative permeability
k_{ro}	=	oil relative permeability
k_{ro}^{*HS}	=	oil endpoint for seawater cycle
k_{ro}^{*LS}	=	oil endpoint when S_{or} becomes constant
k_{rw}	=	water relative permeability
n_1	=	phase Corey's exponent
n_o	=	oil Corey's exponent
$n_o$max	=	maximum typical oil Corey's exponent
N_T	=	trapping number
n_w	=	water Corey's exponent
P_c	=	capillary pressure, critical pressure
P_{cow}	=	oil–water capillary pressure
S_1	=	phase saturation
S_{lr}	=	phase residual saturation
S_{nl}	=	phase normalized saturation
S_o	=	oil saturation
S_{OB}	=	oil bank saturation

S_{orw}	=	waterflood residual oil saturation
S_{or}^{HS}	=	residual oil saturation for seawater cycle
S_{or}^{LS}	=	minimum residual oil saturation by LSWI
S_{wf}	=	water shock front saturation
T_l	=	phase trapping parameter

GREEK LETTERS

β_{Ca}	=	absorbed calcium cation
β_{Mg}	=	absorbed magnesium cation
$\tau^\star$	=	parameter includes heterogeneity and initial oil saturation effects
ω	=	scaling factor
ωS	=	scaling factor for residual oil saturation
λ_{rl}	=	phase relative mobility
θ	=	contact angle, scaling factor
θ^{HS}	=	contact angle for seawater cycle
θ^{LS}	=	contact angle when S_{or} becomes constant
μ	=	chemical potential
σ	=	interfacial tension
ρ	=	phase density

SUBSCRIPTS/SUPERSCRIPTS

high	=	high trapping number
HS	=	high salinity water
i	=	initial
low	=	low trapping number
LS	=	low salinity water
o	=	oil
ow	=	oil-wet
w	=	water
ww	=	water-wet

ABBREVIATIONS

AEC	=	anion exchange capacity
CEC	=	cation exchange capacity
CGI	=	continuous gas injection
CWI	=	carbonated water injection
DoE	=	design of experiment
EOR	=	enhanced oil recovery
EWI	=	engineered water injection
IFT	=	interfacial tension
IOR	=	improved oil recovery
LSWAG	=	low salinity water alternating gas
LSWI	=	low salinity water injection
MED	=	multi-effect distillation
MIE	=	multi-ion exchange
MMP	=	minimum miscibility pressure
MSF	=	multi-stage flash distillation
MVC	=	mechanical vapor compression
NF	=	nano-filtration
NMR	=	nuclear magnetic resonance
OOIC	=	original oil in core
OOIP	=	original oil in place
PV	=	pore volume
PZC	=	point of zero charge
RO	=	reverse osmosis
RSM	=	response surface methodology
SRB	=	sulfate-reducing bacteria
SWAG	=	simultaneous water alternating gas
SWCTT	=	single well chemical tracer test
TCD	=	thermo-compression distillation
TDS	=	total dissolved solids
UTCHEM	=	University of Texas chemical simulator
UTCOMP	=	University of Texas compositional simulator
WAG	=	water alternating gas

APPENDIX C

Dictionary of Geochemical Terms

Activation energy The minimum energy that must be supplied to the reactants in order to start the geochemical reactions and to proceed to the products.

Activity The activity of a certain geochemical specie is the product of its concentration in molality by the activity coefficient of that specie.

Activity coefficient A dimensionless quantity that describes the deviation of a certain geochemical specie from the ideal state.

Activity coefficient models These models express the activity coefficient of a certain geochemical specie as function of both ionic strength of solutions (I_s) and the valence number of that geochemical specie. Examples of activity coefficient models include the Davies model, the extended or WATEQ Debye-Huckel model, and the Setchenow model.

Dissolution/precipitation reactions These reactions are also called solubility product reactions with the solubility product K_{sp}.

Equivalent The molality of a geochemical specie multiplied by the ionic charge of that specie.

Equivalent fraction The equivalent of a geochemical specie on the ion exchange divided by the sum equivalents of the entire ions present on the exchanger. It is also defined as the equivalent of a specie divided by the cation exchange capacity (CEC) of an exchanger because at equilibrium the total equivalents of the ions present on the exchanger is equal to CEC.

Exchange reaction The reaction between geochemical species in the solution and electrolytes attached to a solid surface (exchanger). The ion-exchange reaction is similar to the adsorption phenomena.

Geochemical reaction The process by which some of the geochemical species (reactants) are consumed to produce other geochemical species (products).

Geochemical specie Any chemical entity that is distinguishable from the others by its chemical composition or the phase in which it is present.

Ionic strength A measure of the concentration of ions in that solution which affects the dissociation of geochemical species as well as the solubility of different salts. The definition of ionic strength includes both the concentration of geochemical species in molality and the ionic charge of these species.

Irreversible reaction A geochemical reaction in which the reactants convert to products; however, the products cannot convert back to reactants.

Kinetic reaction A reaction which describes how fast or the rate at which the reaction occurs. This reaction is compared to the local equilibrium reaction. Moreover, usually aqueous-solid reactions including dissolution and precipitation reactions are considered kinetically controlled.

Local equilibrium reaction A reaction which does not depend on the reaction rate and assumed to be at equilibrium. Usually, homogeneous (aqueous) reactions are considered at equilibrium. Moreover, heterogeneous reactions (dissolution/precipitation) can be treated at equilibrium if the flow rate is low.

Molality A unit of expressing the concentration of geochemical species and is defined as the amount of substance (in moles) in one kilogram of solvent.

Molarity A unit of expressing the concentration of geochemical species and is defined as the amount of substance (in moles) in one liter of solution.

pH A measure of the concentration of hydrogen ion in moles per liter of solution (molarity) and is defined as the negative of the logarithm of hydrogen activity in molarity.

Redox reaction A reduction–oxidation reaction that occurs between the same elements, but with different oxidization states or valences.

Reversible reaction A geochemical reaction in which reactants convert to products and vice versa. In this reaction, reactants and products are never fully consumed.

Saturation index (SI) An index which states the saturation status of a solid geochemical specie in the geochemical system. It is defined as the logarithm of the ratio of the ion activity product (IAP) of a solid specie divided by the solubility product (K_{sp}) of that solid species. This index indicates whether solid geochemical species are at equilibrium, undersaturated or supersaturated with the solution.

Super/undersaturation A positive saturation index of a solid indicates a solution that is supersaturated with that solid and the solid is precipitated in the system. A negative saturation index of a solid specie indicates a solution that is undersaturated with respect to that solid and in case that solid is present in the system, then it dissolves to the aqueous phase.

Surface complexation A general form of ion-exchange reaction in which the species are attached to an existing functional group of the solid surface including amorphous aluminosilicates, metal oxides/hydroxides, and organic matters. Usually, the ion-exchange reaction concept is applied on major ions whereas the surface complexation reaction is used for minor and trace components. In surface complexation reactions, the electrical potential of the surface is considered in the sorption of geochemical species from the surface.

INDEX

Note: Page numbers followed by "*f*" and "*t*" refer to figures and tables, respectively.

M

N

O

Kolibri 360 GmbH
Ferdinand-Jühlke-Straße 7
99095 Erfurt, Deutschland
produktsicherheit@kolibri360.de